WATER: IMPACT OF DAMS ON ITS QUALITIES

WATER

Impact of Dams on its Qualities

Editor

Bharat Jhunjhunwala

KALPAZ

ISBN: 978-81-7835-997-7

Price : ₹ 695

First Published 2014

Published by

Kalpaz Publications
C-30, Satyawati Nagar,
Delhi - 110052
E-mail: kalpaz@hotmail.com Ph.: 9212142040
Laser Typesetting : Shah Computer Graphics, Delhi
Printed at : G. Print Process, Delhi

Cataloging in Publication Data--DK

Courtesy: D.K. Agencies (P) Ltd. <docinfo@dkagencies.com>

Jhunjhunwala, Bharat, 1950-
Water : impact of dams on its qualities / Bharat Jhunjhunwala.
p. cm.
Includes bibliographical references.
ISBN 9788178359977

1. Dams--Environmental aspects--Ganges River (India and Bangladesh) 2. Water quality--Ganges River (India and Bangladesh)---Effect of dams on. 3. Ganges River (India and Bangladesh)--Environmental conditions. I. Title.

DDC 577.2720954 23

For

Swami Gyan Swarup Sanand

Contents

PART 3: IMPACT OF DAMS ON PHYSICAL AND CHEMICAL QUALITIES OF WATER

PART 4: IMPACT OF DAMS ON BIOTIC QUALITY OF WATER

PART 5: CONCLUSION: ALTERNATIVE WAYS OF GENERATING HYDROPOWER

Abbreviations

AHEC	Alternate Hydro Energy Centre
APS	American Physical Society
BOD	Biological Oxygen Demand
CDM	Clean Development Mechanism
Cond	Conductivity
COPPE	Engineering Department, Federal University of Rio de Janeiro
CPCB	Central Pollution Control Board
DO	Dissolved Oxygen
GHG	Green House Gasses
GWh	Giga watt hours
HEP	Hydro Electric Project
IIT	Indian Institute of Technology
INPA	National Institute for Research in the Amazon
IPCC	Intergovernmental Protocol on Climate Change
IRN	International Rivers Network
kWh	Kilo watt hours
MRET	Molecular Resonance Effect Technology

MW	Mega Watt
NEERI	National Environment Engineering Research Institute
NHPC	National Hydro Power Corporation
NMR	Nuclear Magnetic Resonance
NPC	National Productivity Council
NTU	Nephelometric Turbidity Units
ORP	Oxidation-Reduction Potential
PSI	People's Science Institute
SWI	Shannon Weiner Index
TDS	Total Dissolved Solids
TERI	The Energy Research Institute
THDC	Tehri Hydro Development Corporation
TVA	Tennessee Valley Authority
UV	Ultra Violet
WQI	Water Quality Index
WTP	Willingness to Pay

Preface

I am an economist by training. I started living on the banks of River Alaknanda in 2003 so as to be in solitude and pursue my spiritual growth. Things were to be otherwise. Soon, in 2007, the Government proposed to build the Kotlibhel 1B hydropower project which would convert the free-flowing and singing Alaknanda into a reservoir. People of my village as well as Vimal Bhai came to me to help prevent this project. Thereafter, I was slowly sucked into the 'Save Ganga from Hydropower' movement.

I was a supporter of big dams like Tehri till then. Detailed study of the Environment Impact Assessment of Kotlibhel 1B project convinced me that these hydropower projects did not contribute to economic growth and welfare of the people. The environmental costs were much greater than the benefits from generation of electricity.

One argument proved difficult to defend. This was regarding the spiritual powers of the Ganga. There seemed to exist no credible evidence that Ganga had any spiritual powers and that they were harmed by hydropower projects. I decided to write this book to fill in that gap.

As I said, I am an economist by training though I have studied some Chemistry and Physics during my Bachelor's

Degree. I beg the reader to be kind and ignore my shortcomings in this venture into an area that I am not competent to enter into.

Bharat Jhunjhunwala
Lakshmoli, PO Maletha, Kirti Nagar,
Uttarakhand 249161 India
bharatjj@gmail.com
+91-85278-29777
January 14, 2013

Acknowledgements

I thank all the contributors to this book for writing papers especially for this book and for granting permission to use their works.

I thank Ganga Aavahan for providing the photographs of water crystals of the Bhagirathi upstream and downstream the Tehri project.

I thank my wife Madhu for bearing through my activism and providing me emotional and homely support always.

List of Contributors

Swami Shantidharmananda, Swamiji lives on the banks of the Ganga in Rishikesh. He was an engineer with Hindustan Aeronautics Limited before taking Sannyasa. He teaches Hindu scriptures.

Dr. Zakia Akhtar, is Lecturer (School Education), Jamia Millia Islamia, New Delhi.

Fr Subhash Anand, St. Paul's School, Bhupalpura, Udaipur.

Bharat Jhunjhunwala, is a Columnist. He was formerly Assistant Professor of Economics at Indian Institute of Management, Bengaluru, India.

Tiziano Paolini, Aqua Viva, Switzerland.

Dean Radin, Institute of Noetic Sciences, Petaluma, California, USA.

Gail Hayssen, Institute of Noetic Sciences, Petaluma, California, USA.

Masaru Emoto, IHM Research Institute, Tokyo, Japan.

Takashige Kizu, IHM Research Institute, Tokyo, Japan.

G.D. Agrwal, Since taking Sannyasa he is known as Swami Gyan Swarup Sanand. He is Former Professor, Civil

Engineering, IIT, Kanpur and Member Secretary, Central Pollution Control Board.

International Rivers Network is a NGO based in California, USA dedicated to conservation of rivers worldwide.

Penny Schmitt, U.S. Army Corps of Engineers.

R.C. Sharma, Department of Environmental Sciences, HNB University, Srinagar, Uttarakhand.

Punam Chauhan, Department of Environmental Sciences, HNB University, Srinagar, Uttarakhand.

Manju Bahuguna, Department of Environmental Sciences, HNB University, Srinagar, Uttarakhand.

Anil Gautam, People's Science Institute, Dehradun.

Paritosh Tyagi, Formerly Chairman, Central Pollution Control Board.

Steve Gamble, is a healer and complementary medicine Practitioner, Researcher and Author.

Craig Hoffman, AquaLiv Inc.

Alex Andersson, Author of 'Living Water.'

Wikipedia.

Andy Coghlan, Reporter with New Scientist.

Dan Stewart, Canada.

Denise Routledge, Canada.

Howard Fisher, Natural Anti-Aging Physician, is the founder and operator of the Glen Park Clinic, Toronto.

Introduction

I have been agitating against the construction of hydropower projects on the River Ganga. Argument made consistently by proponents of these projects is that the projects are not damaging the water in any way. Most projects being made on the Ganga do no envisage storage of water for long durations. Water received during the course of the day is released in the same day though at different times. These projects store water during the night and day and release during morning and evening to generate 'peaking' power.[1] No water is consumed during generation of electricity. The project proponents say that making the projects is a win-win situation. Water remains intact while electricity is generated additionally.

These projects are of two types–tunnel– and reservoir based. A barrage is made across the river bed in tunnel-based projects. Water is diverted into a tunnel and taken to the power house downstream. Water is released back into the river after generation of electricity. A bigger dam is made across the river bed in reservoir-based projects. A long reservoir—sometimes extending up to 40 km—is made behind the dam. Water is taken

1. These projects are called 'Run-of-the-River' projects. However, this is a misnomer because run of the river is obstructed and water flows only during certain hours of the day.

from the upper level of the dam, passed through turbines and released back into the river. Flow of the river is obstructed in both types of projects. Water is made to flow through the tunnels depriving it of contact with air and earth in tunnel-based projects. Water is held in reservoirs for long duration in reservoir-based projects. The velocity is reduced drastically and friction with stones is removed altogether. In both cases the flow of river water is modified.

The project proponents argue that this change in flow does not impact the quality of water. This book has been assembled to shed light on the fact that hydropower projects deeply impact the quality of water. This impact is made on the physio-chemical, biotic and subtle qualities of water.

The argument in favour of harnessing the rivers for generation of hydropower is that of economic growth. It is said that we need electricity for economic growth and uplifting the living standard of our peoples. It is assumed that the spiritual qualities of river waters, if any, are either not impacted by the hydropower projects; or there is no negative economic impact if that indeed takes place. Therefore, we need to harness the rivers for economic growth which is the greatest need today.

There are two ways in which river water impacts economic growth. First way is through generation of electricity. Second way is through the psychic impact on the pilgrim which leads to unleashing the pent-up energies in his psyche. The net impact of river on economic growth depends on the sum of these two impacts.

The psychic impact is dependent on the fact whether river water, especially of the Ganga, has any psychic or spiritual qualities. I have come to believe that waters of the Ganga have special spiritual and psychic qualities; and that these are irreparably harmed by making the water flow through tunnels, reservoirs and turbines. The loss of psychic powers of the river waters will negatively impact the economic growth because the psychic energy will not be released. Economic growth will be negatively impacted by hydropower projects via the loss of psychic qualities. I make this argument only to show that hydropower is not a desirable activity even if economic growth is considered to be the main objective of society.

I do not agree with putting economic growth at the centre of our life. Economic growth is desirable only because it provides satisfaction, happiness or welfare to the people. I believe that happiness of our people is dependent on their psychic makeup. This psychic makeup is deeply influenced by taking a dip in the Ganga. Therefore, it may be desirable to give up some economic growth by not making hydropower. The gain to the people's happiness from taking dip in the Ganga may be much more than the loss of happiness from less generation of electricity.

I argue that we must use the 'higher' spiritual qualities of river waters and, if necessary give up the 'lower' physical quality of gradient and flow that is used for generation of hydropower. India's destiny is to contribute in the spiritual domain. We should not destroy that in the mistaken belief that waters of the Ganga do not have any special psychic powers.

An alternate model of economic development would be to establish universities, hospitals, design centres and software parks along the banks of the Ganga. These activities, I suggest, will be more successful on banks of the Ganga than elsewhere because the psyche of the workers will be more balanced and active. The services sector comprising of these activities has, in any case, become the engine of growth of our economy. Use of the psychic powers of the Ganga will give a boost to the services sector and more than compensate for the loss of growth due to abstaining from hydropower.

This formulation is critically dependent on the belief that water has the capability to carry spiritual or psychic charges; and that Ganga has special charges that are harmed by hydropower projects. This book has been written to explore this dimension.

As a second best alternative, is collaboration with Shri Paritosh Tyagi, former Chairman, Central Pollution Control Board, I suggest that part of the water of the Ganga may be drawn for generation of hydropower. This water may not be put back into the Ganga. It may be taken to the plains through a cascade of hydropower dams and put in the canals for use in irrigation. Part of the water that is allowed to flow freely will then carry the psychic charges and we shall not be wholly deprived of that benefit.

This book is structured as follows:

In first section I give evidence that water is used as a medium of spiritual activities in the three major religions of the world—Hinduism, Islam and Christianity. This is to show that the demand for preserving the spiritual qualities of water is not a Hindu parochial thought but is common to these religions.

In second section I provide whatever evidence is available to show that water of rivers do carry spiritual charges and that these are damaged by hydropower projects.

In the third section I provide evidence that hydropower projects make harm to the physio-chemical qualities of river waters.

In the fourth section I provide evidence that hydropower projects make harm to the biotic qualities of river waters. A physically healthy river water alone will be able to carry the psychic charges.

In the fifth section a quick discussion of alternative method of generating hydropower is described so that we generate some hydropower and also conserve the spiritual qualities of the Ganga.

In the sixth section I provide additional evidence for the psychic qualities of water in general. This is to bolster the argument that river water can and does carry spiritual charges.

PART 1: WATER IN RELIGIONS

Introduction

The purpose of writing this book is to place before the reader the available scientific evidence indicating that the Holy Ganga River may have some spiritual qualities that will be irreparably destroyed by making hydropower projects on the river.

Hindus believe that water of the Ganga provides salvation. There are seven psychic centres in the spine of a human being called chakras. A person's mind is held captive by the impulses emanating from these chakras. These impulses may be rooted in one's past lives or in the earlier experiences of the present life. The origin of these impulses is not relevant for the present discussion. One is considered to have attained salvation if the psychic signals emanating from these chakras come to an end. Then a person becomes 'free'.

The Ganga flows below the Temples of Badrinath, Kedarnath and Gangotri. These temples contain spiritual charges. These charges are absorbed by the Ganga as She flows below these Temples. The person taking a dip in the Ganga is unconsciously influenced by these charges. Such a dip creates a direct connection between the psyche of the person and the psychic charges of these temples. In this way the psyche of the person is regularised, smoothened, quietened and made simple and easy. This is salvation.

Now, if waters of the Ganga indeed have such a capability of carrying spiritual charges as believed by the Hindus, then it is useful to examine if other religions also endorse the existence of such spiritual qualities of water.

In this section I present three articles on the spiritual qualities of water from the Hindu, Islamic and Christian perspectives. It is shown that there are deep traditions in all these major religions of the world based on the spiritual qualities of water. Thus the belief that waters can be a medium of spiritual development is not an idle imagination. It is endorsed by all these religions.

1

Water in Hinduism

Swami Shantidharmananda

Editor's Introduction: The writer explains that water has the capacity to absorb spiritual charges. The spiritual content of water depends upon the penance and other practices performed on its banks. River performs the function of absorbing these spiritual charges at places where penance is being done; and delivering these charges to the pilgrims when they take a dip in the river. Water is the medium through which spiritual charges are easily taken by the pilgrim.

Water has 'lower' physical as well as 'higher' spiritual qualities. By implication, we should conserve, honour and use the higher spiritual qualities and not destroy them for exploitation of its lower physical qualities.

The author also writes that river water longs for contact with the air. Deprivation of this contact may make the water 'numb' or 'dead'.

Nature of Water

Holy Water has been used in religious rites by the Hindus, Buddhists, Jains, Christians, Muslims and other religions since ancient times. How does water become Holy? The Sages, Saints, Monks, Priests and Spiritual Practitioners undertake penance, meditation, fast, chanting of sacred books, repetition of mantras orally or mentally, observe silence, go into seclusion and other spiritual and religious practices. These practices create subtle vibrations that can be captured by an equally subtle receptor. Thus a monk can communicate with his disciple while remaining totally silent. Water has the property of receiving and capturing and holding these vibrations. These vibrations are transferred to the person who touches or drinks that water. In this way water becomes a medium to transfer and spread the spiritual vibrations.

I had gone to Singapore in 2006 to attend the inauguration of a yoga center. I saw that Buddhist monks of two different sects consecrated a bowl of water by chanting some mantras. Then they asked me to consecrate the same water by chanting Vedic mantras, which I did, and gave back that bowl of Holy Water to them. Then chanting some other mantras they sprinkled that holy water on all the devotees who had assembled over there for the inaugural function of the yoga hall. They also sprinkled that holy water in the whole building including kitchen, etc. In this way that ordinary building was transformed into a temple-like spiritual centre conducive for yogic practices.

The spiritual energy can be transferred to water in other ways as well. Dissolving different materials in water can change its spiritual content. For example, sandalwood has certain vibrations. Dipping a stick of sandalwood will pass its charge to water. Or, pouring water on an idol will create similar charge in the water.

Let us say we are worshipping a deity who is destroyer of sins. Now, water is consecrated by chanting mantras and then it is poured on the idol. The charge of 'destroying the sin' will get transferred to the water. The worshipper who partakes of that water will have some of his sins removed. Here 'sin' must be thought of as particular tendency or vibration in the spinal column. Partaking of the holy water will change those vibrations in the direction of the vibrations emanating from the idol. I have seen this done since my childhood in my house, temples and other religious places that I have been to in India and abroad.

I believe that water can carry and preserve these charges even after it is removed from the source of energy. Thus, devotees carry water of the Ganga River to their home. It is due to this holding quality of water that religious rites are performed by water in most religions.

Origin of Life

Water is the very foundation of life on earth, and also it is the cause of transformation of every living being from one birth to another birth. Thus the origin of life is water. This is due to its ability to conserve and preserve information even after the source of information is removed or destroyed. In this regard thus states the Chandogya Upanishad –

वेत्थ यथा सौ पंचम्यामपहुतावाप: पुरुावचसो भवन्तीति (5.3.2).
तस्मातस्मिाग्नौ देवा रेतो जुह्वति तस्या आहुतेर्गर्भ: संभवति (5.8.2).

The master asks, "Do you know how Water comes to be known by the word 'Person' after the fifth oblation."

He then relpies, "(Woman is indeed the fifth fire.) The gods offer semen in that fire. From which the foetus comes into being which is later known as Person."

Semen is a form of water. Thus birth of a being takes place through water.

The complete procedure is as follows. A person performs religious rites. Say, he offers oblations with fire. The oblations enter the intermediate zone of space as smoke, flame and sparks of the sacrifice. They transform the intermediate space into particular vibrations which carry the wishes and resolutions of the person's mind. The vibrations in space then interact with the mind of other living beings on earth and the thoughts of the worshipper influence the people of the entire area. In this way the sacrifice gives him the result of the sacrifice.

Now water is used in the sacrifice in various ways—sprinkled directly as well as through water contained in fruits and grasses. The vibrations created in space are called 'Water' because of their association with water while the rites were performed.

The vibrations return to the earth in form of rain and enter crops like paddy, wheat, barley, vegetables, spices, etc. These are eaten by human beings and are transformed into semen, which enters a woman's womb. There the oblations take a human form.

In this way water carries the thoughts and resolutions of human beings. Thoughts of man enter the sacrifice through grains and wood used in sacrifice. They pass into space during the sacrifice. From space they return to earth via rain and enter the crops again. Human beings eat the crops and those vibrations enter the semen and are reborn through a woman's womb.

The Five Fires

The cycle of movement from earth to space and back is explained in the third section of the fifth chapter of Chandogya Upanishad in the form of five fires.

The First Fire. The sacrificer takes a sankalpah or vow or commitment or a resolution with water in his palm. This water imbibes these thoughts. Thoughts of the sacrificer or the deceased enter the Intermediate Space along with the oblations made during the sacrifice. Likewise, the water in the body of the deceased imbibes the thoughts of the deceased and thoughts of the deceased enter the Intermediate Space upon cremation. The Intermediate Space or Heaven is the first 'fire'. The Sun is the 'fuel' of this first fire since the Intermediate Space shines by the Sun. Therefore, the Sun is called 'fuel'. The rays are the 'smoke' because they emerge from the sun, just as smoke rises from wood. The day is the 'flame' because of the similarity of luminosity and it being brought into being by the Sun. The moon is the 'ember.' Embers appear after the flame subsides just as moon appears after the sun sets. The stars are the 'sparks' because they are scattered all over. The sacrificer offers oblations in such fire in the form of water. Cloud is the result of the first oblation.

The Second Fire. The cloud is the second fire. Air is the fuel of the cloud, the haze is the smoke, lightning is the flame, thunder is the ember, rumbling of the clouds are the sparks. In that fire the sacrificer offers oblation. Rain is the result of that oblation.

The Third Fire. The Earth is the third fire. Time is the fuel, sky is the smoke, night is the flame, directions are the embers and intermediate directions are the sparks. Rainfall is the oblation which carries the vibrations of the sacrifice. Food is produced from earth.

The Fourth Fire. Man is indeed the fourth fire. Speech is the fuel, outgoing breath is the smoke, tongue is the flame, eyes are the ember and ear is the spark. In that fire the gods offer food as oblation. From that oblation is created semen.

The Fifth Fire. Woman is the fifth fire. The genital organ is the fuel, that which he (the husband) beckons is the smoke, the vagina is the flame, the husband's genital organ which is inserted is ember and the bliss experienced by both is the spark. In that fire semen is the oblation. From that oblation the foetus comes into being, which results in the birth of a child.

Thus, through the stages of cloud, rain, food and semen, water comes to be called a 'person' after the fifth oblation. This cycle of rebirth is based on water.

Sacredness of the Ganga

I have shown above that water carries the thoughts of the sacrificer. This happens more directly with the flowing river. Water of the river carries the thoughts of those on her banks. The vibrations of the rituals performed on its banks and the penance observed by sages enters the water. Therefore, sacredness of a river depends on the rituals and penances performed on her banks. All rivers are not equally sacred since worship is not done on their banks as is the case with Ganga, Yamuna, Mandakini and Alaknanda.

Ganga Dashahara brings throngs of bathers to the riverfront in Haridwar and all Tirthas (Pilgrimage places) on the dashami (tenth day) of the waxing moon fortnight of the lunar month of Jyestha (June approx). Hindus celebrate the *avatarana* or descent of the Ganges from heaven to earth on this day. A dip in the Ganges on this day is said to rid the bather of ten sins. This belief is the origin of the name Dashahara: dasha = ten; hara = 'to destroy'. This is also interpreted as destruction of sins of past ten lives. Those who cannot journey to the river, however, can achieve the similar result by bathing in any nearby body of water, which, for the true believer in the Hindu tradition, takes on most of the attributes of the Ganges on this particular day due to invocation of the Ganga.

These beliefs of the Hindus are all based on the spiritual qualities of waters of the Ganga.

Redemption by Ganga

The photograph at the Manikarnika Ghat at Varanasi has the following caption, "Who dies in the waters of the Ganges obtains heaven."

Since Ganga had descended from heaven to earth, she is also the vehicle of *ascent* from earth to heaven. One of her names is *Triloka-patha-gamini.* This means 'one who travels to the three worlds.'[1] Consequently Ganga is considered to be a "tirtha," or crossing point, for all living beings as well as the dead irrespective of the world in which their soul is located at present.

Those who are lucky to die in Varanasi and are cremated on the banks of the Ganga are granted instant salvation. Here salvation must be understood as removal of desires that cause rebirth. The unfulfilled desires of the deceased are carried by the waters of his body to the Intermediate Zone as explained above. Cremation on the Banks of Ganga at Varanasi influences these psychic tendencies and smoothens or removes them. Thus the soul of the deceased becomes free of desires and is freed from the cycle of rebirth.

If the death has occurred elsewhere, salvation can be achieved by immersing the ashes in the Ganges because the soul remains in close contact with the ashes. If the ashes have been immersed in another body of water, a relative can still gain salvation for the deceased by journeying to the Ganges, if possible during the lunar "fortnight of the ancestors" in the Hindu calendar month of Ashwin (September approx) and performing the *Shraddha* (offerings with special faith) rites. Tradition holds that souls of the ancestors are 'open' and looking towards the earth in this period.

Ganges is so important in the rituals after death. The *Mahabharata*, in one of its popular *shlokas*, says, "If only (one) bone of a (deceased) person should touch the water of the Ganges, that person shall dwell honoured in heaven." *Kashi Khanda* of the Skanda Purana recounts the remarkable story of *Vahika*, a profligate and unrepentant sinner, who is killed by a tiger in the forest. His soul arrives before Yama, the Lord of Death, to be judged for determining his fate in the life hereafter. Vahika's soul is at once dispatched to hell since he had no virtue. While this is happening, his body on earth is being picked by vultures, one of whom flies away with a foot bone. Another bird comes after the vulture, and in fighting him off, the vulture accidentally drops the bone into the Ganges below. Blessed by this happenstance, Vahika, on his way to hell, is rescued by a celestial chariot which takes him instead to heaven.[2]

Special Qualities of Water

The spiritual importance of the rivers is anchored on water being a medium for transfer of spiritual charges. A person may stand before an idol and may not be able to connect with the spiritual charges in the idol. The same connection can be made more easily by pouring water on the idol and then the person partaking those waters. Rivers are important, because they flow, i.e., they carry the charges to different places and peoples. Ganga is especially important among the rivers because the charges are more intense. The ability of water to carry charges is the rockbed of this concept.

Water taken by a living being gets divided into three parts– Subtle, Medium and Gross. The Chandogya Upanishad states:

> आपः पीतास्त्रेधा विधीयन्ते। तासां यः स्थविठो धातुस्तन्मूत्रं भवति।
> यो मध्यमस्तल्लोहितं। योणिठः स प्राणः। (6.5.2)

> Water that we drink gets divided into three parts. Of it, that which is the grossest ingredient becomes urine, that which is the medium constituent becomes blood and that which is the subtlest becomes the vital force.

The conversion of water into vital force depends upon the vital force present in the water. The vital force in water is dependent upon the energy absorbed by it from nature. It collects energy while it is continuously moving, swaying from side to side, bending and curving, spiralling and flowing over rocks and absorbing different minerals. During its natural flow, water of a river absorbs energy, carries energy and emits energy. Thus drinking as well as doing worship with flowing water is recommended.

Water emerges from mountains and flows through mountain valleys, enters the plains and continues its journey to the ocean. It flows in a zigzag way, playing with the wind and dashing against the rocks. In this process it captures vital energy. On the other hand, when water is confined into a lake or and forced to flow through a restrictive system like tunnel, it will lose its energy. Ganga maintains its natural characteristics through contact with its source of energy. Polluting substances or other negative energies do not contaminate it much due to the great amount of worship and prayer of which it is recipient.

Water Memory

Water is a carrier of information and also it preserves information even after the source of information is removed or destroyed. It is due to this quality of water that Vedic, Hindu and many non-Hindu religious rites are performed by water.

Water will lose this vital force if it is made to flow in tunnels and pipes. In this regard Vedas say:

मैथुनेच्छावतीनां तरंगिनीनामपां वायुः पठेत्यवर्तत ततो वामदेव्यं साम अभवत्।

Flowing water wishing for intercourse was overtaken by wind fulfilling its wish due to which a Sama called Vamadevyam came into existence.[3]

This verse indicates that water has a different level of energy when it is in playful contact with air. Water will be deprived of this contact with air if it is made to flow through tunnels.

Conclusion

Water has two levels of existence. The physical level is sought to be exploited in generation of hydropower by construction of dams. But this will lead to the destruction of its vital force, which, in turn will lead to less energy among the people. So both the 'lower' physical and 'higher' vital levels of existence must be kept in mind before taking a decision to generate hydropower from our rivers—especially the Ganga.

Notes

1. *triloka*= "three worlds"; *patha* = "road"; *gamini* = "one who travels".
2. http://en.wikipedia.org/wiki/Ganges.
3. See Bhashyam and Anandagiri commentaries on Chandogya Upanisad 1.13.1 & 2.13.1.

2

Water in Islam

Zakia Akhtar

Editor's Introduction: The author quotes from the Holy Quran to establish that Paradise has *flowing* rivers in contradistinction to stagnant lakes or pools. This indicates that flow is an essential element of holy water.

The author tells us about the miraculous qualities of Zamzam water which flows from a well near the Kaaba. This water is believed to have properties of improving learning as well as health. This indicates that water has certain quality that enables it to absorb and transfer spiritual charges.

Islam is a universal religion born in the harsh deserts of Arabia to complete the message of former prophets and to convey the divine revelation in its last testament (Qur'an). Islam ascribes the most sacred qualities to water as a life-giving, sustaining, and purifying resource.

It is the origin of all life on earth and the substance from which God created man (Qur'an 25:54).

"We made from water every living thing" (Qur'an 21:30).

In the coming paragraphs we will try to analyse the relation between Islam and water with some examples.

Water in the Quran

The word (Mâ) means water. It occurs more than sixty times in the Quran. The Quran deals with water both on symbolic and practical levels. It is said that water is the primary element that existed even before the heavens and the earth did:

> And it is He who created the heavens and the earth in six days, and his Throne was upon water (Qur'an 11:7).

The water of rain, rivers and fountains runs through the pages of the Qur'an to symbolise God's benevolence:

> He sends down saving rain for them when they have lost all hope and spreads abroad His mercy (Qur'an 25:48).

The believers are constantly reminded that it is God Who gives sweet water to them and that He can just as easily withhold it:

> Consider the water which you drink. Was it you that brought it down from the rain cloud or We? If We had pleased, We could make it bitter (Qur'an 56:68-70).

In this verse, the believers have been warned that they are only the guardians of God's creation on earth; they must not take His law into their own hands.

Water is important both for cleansing and for purifying. Muslims must be ritually pure before approaching God in prayer. Some mosques have a courtyard with a pool of clear water in the centre, but in most mosques, the ablutions are done outside the walls. Fountains symbolising purity are also sometimes found in mosques. In Islam ritual purity (called *tahaarat*) is required before carrying out religious duties especially salat (worship).

There are three kinds of ablutions. Firstly, *ghusl*, the major ablution, is the washing of the whole body in pure water, after declaring one's intention to do so. Muslims are obliged to perform *ghusl* after sex, which incurs a state of major impurity. *Ghusl* is also recommended before the Friday prayer, the two main feasts, and before touching the Quran. *Ghusl* is also an obligatory ritual for the dead before they are buried.

The second ablution is *Wudu*. It is a minor ablution, which is performed to remove minor ritual impurity from everyday life. This must be done before each of the five daily prayers and involves using pure water to wash the face, to rub the head and to wash the hands and arms up to the elbows and the feet up to the ankles. This comes from the Quran:

> O you, who believe, when you prepare for prayer, wash your faces and your hand to the elbows; rub your head and your feet to the ankles (5: 7-8).

This is elaborated in detail in the Sunna. Every mosque has running water for *Wudu*.

The third type of ablution is performed when no water is available. In this case, clean sand may be used. In addition, water is also used in some rituals on some special occasions.

Symbolism of Water as an Element of Paradise

Water is used as a symbol of resurrection since Paradise is always described as a place with rivers flowing and florid vegetation (Qur'an16: 30-31).

Metaphors in which water is used to symbolize Paradise, righteousness, and God's mercy occur quite frequently in the Quran. There are numerous references to cooling rivers, fresh rain and fountains of flavoured drinking water in Paradise. We can deduce from these that water is the essence of the Garden of Paradise. It flows beneath and through the Garden, bringing coolness and greenery, and quenching thirst. It is said that:

> (The believers will be rewarded for their piety by) rivers of unstagnant water; and rivers of milk unchanging in taste, and rivers of wine, delicious to the drinkers, and rivers of honey purified (Qur'an 47:15).

Water in Paradise is never stagnant; it flows and rushes, unlike the festering waters of Hell. This is especially important in the context of the policy of the Indian Government to harness the rivers for the generation of hydropower. The flowing rivers will be converted into stagnant reservoirs. In other words, Paradise will be converted into Hell.

The Qur'an also equates the waters of Paradise with moral uprightness: "In the garden is no idle talk; there is a gushing fountain" (Qur'an 88:11-12).

The specific statements about Paradise in the Qur'an have led to attempts to map the Paradise. Muslim rulers from Moorish Spain to Persia sought to reproduce the image of Paradise in the design of their palace gardens, creating elaborate water features, pools, and fountains. The gardens of the Alhambra in Spanish Granada, the Bagh-é-Tarikhi in Iran's Kashan, and the gardens of the imperial palaces in Morocco's Marrakesh all testify to this desire to emulate Quranic Paradise on earth. All are designed around water features and fountains that have been subtly woven into the layout of the beautiful parks, hence combining flowing water and the beauty of natural landscape to fill the human soul with faith, joy and happiness.

Water in Islamic Society

Water is an integral part of Islam and Islamic society. It is frequently found in Islamic literature be it an account of a water body or a poem set in an oasis. We can also see its impact on art and architecture in Islamic civilisation. In earlier times the city was built around a mosque where water was present for ablution and city streets had *sabils* (public fountains built for charity). If we see the historical structures of Islamic cities, we will get to know that those had Hammam (Turkish bath).

Water also influenced Islamic iconography. One can see calligraphic inscriptions with verses from Quran related to water near ablution pools or *sabils*.

These examples show that water and Islam are intertwined. In this link we cannot forget the very important and sacred Zamzam water.

The Sacred Waters of Zamzam

The Water of Zamzam comes from a well located within the Majid-al-Haram in Mecca, Saudi Arabia, 20 m east of the Ka'ba, the holiest place in Islam. History of Islam states that the Zamzam Well was revealed to Hajara, the second wife of Prophet Abraham and mother of Ismail. According to Islamic tradition, she was desperately seeking water for her infant son but she could not find any because Mecca is located in a hot dry valley with few sources of water. Then Hajara ran seven times back and forth between the two hills of Safa and Marwah, looking for water. Getting thirstier by the second, the baby Ismail scraped the land with his feet, where suddenly water sprang out.

The name of the well comes from the phrase Zomë Zomë, meaning 'stop', which, according to a legend, was a command repeated by Hajara during her attempt to contain the spring water.

According to Islamic belief, it is a miraculously generated water that has been granted by Allah. Millions of pilgrims visit the well each year while performing the Hajj or Umrah pilgrimages in order to drink its water. Muslims also believe that the Zamzam well is a miracle. It has never gone dry despite the millions of litres of water of the Well being consumed every year. It had been deepened several times in history during periods of severe droughts.

The Zamzam well was excavated by hand, and is about 30 m deep and 1.08 to 2.66 m in diameter. Originally, water from the Well was drawn via ropes and buckets. Nowadays the Well is located in a basement room where it can be seen behind glass panels but entry is not allowed for pilgrims. Electric pumps draw the water, which is available throughout the Masjid al-Haram via water fountains and dispensing containers near the Tawaf area.

Hydrologically, the well is in the Wadi Ibrahim—Valley of Abraham. The upper half of the well is in the sandy alluvium of the valley, lined with stone masonry except for the top 1 m, which has a concrete "collar". The lower half is in the bedrock.

Between the alluvium and the bedrock is a section of permeable weathered rock, lined with stone, and it is this section that

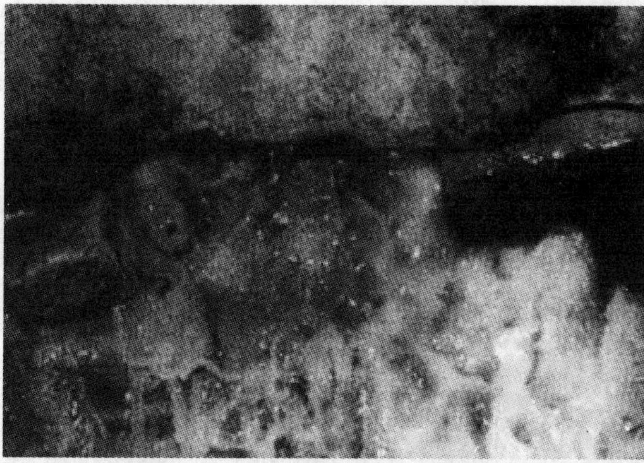

Plate 1: The source of Zamzam Water.

provides entry for the main water into the well. Water in the well comes from absorbed rainfall in the Wadi Ibrahim, as well as run-off from the local hills. Since the area has become more and more settled, water from absorbed rainfall on the Wadi Ibrahim has decreased.

The well had two cisterns in the first era. One was for drinking and one for ablution or Wudu. At that time, it was a simple well surrounded by a fence of stones. Then in 771 A.D. (149 A.H.) a dome was built above the well, tiled with marble. In 775 A.D. (153 A.H.), the well was rebuilt with a dome of teak which was covered with mosaic. One small dome covered the well, and a larger dome covered the room for the pilgrims. In 835 A.D. (213 AH) there was further restoration, and the dome was covered with marble.

In 1417 A.D. (795 A.H.), the mosque was damaged by fire, and required restoration. Further restoration was done in 1430 A.D. (808 A.H.), and again in 1499 A.D. (877 A.H.) when the marble was replaced.

The most extensive restoration was done to the dome in 1915 A.D. (1333 AH). The building housing the Zamzam was moved away from its original location to get it out of the way of the Tawaf when millions of pilgrims would circumambulate the Ka'ba. The water of the well is now pumped to the eastern part of the mosque where it is made available in separate locations for men and women.

Plate 2: Pilgrims near Zamzam Well.

The Process of Making Zamzam Water Sacred

In almost all religions water is considered as divinely blessed with positive energy. Some sects have the notion that humans themselves help charge it. Therefore, with millions of people making the journey to the Zamzam well every year you can imagine that this water has been blessed and charged with positive energy many times over. Muslims all over the world find this to be the sole root of Zamzam's power. It is a human construct, which provides it a distinct identity.

In this context, I remember one of my childhood memories when my grandmother was on deathbed and my mother dropped the Zamzam water in her mouth. This was not the only incidence where Zamzam water was used. In fact, most of the Muslims keep this water in their homes and use it especially at the time of sickness. Apart from its physical properties, this sacred water has some subtle properties, which makes it different.

One of the miracles of Zamzam water is its ability to satisfy both thirst and hunger. According to Prophet Muhammad S.A.W. (the last Prophet), the best water on the face of the earth is the water of Zamzam; it is a kind of food and a healing from sickness (*Saheeh-al-Jami*, 3302).

It is also said that Prophet Mohammad also used to carry Zamzam water in pitchers and water skins back to Medina. He used to sprinkle it over the sick and make them drink it (*Al-Silsilat al-Saheehah,* 883).

The results of tests conducted in European laboratories and the Ministry of Agriculture and Water Resources, Saudi Arabia revealed that the quantity of calcium and magnesium salts was slightly higher in Zamzam water (Hussain 2010). This may be why this water refreshes tired Hajis, but more significantly, the water contains fluorides that have an effective germicidal action. Zamzam water's appeal is universal. This water has never been chemically treated or chlorinated, as is the case with water pumped into the cities. Biological growth and vegetation usually takes place in most wells. This makes the water unpalatable owing to the growth of algae causing taste and odour problems. However, in the case of the Zamzam water well, there was no sign of biological growth (Griffin 2007).

There are plenty of examples showing that Zamzam water has healed the incurable diseases. One case is related to Talal, the younger brother of Kuwaiti journalist Ghneim al-Muteyri who was suffering with leukaemia. American doctors in Riyadh performed a spinal cord transplant on him. Talal spent several months taking various medications. However, they did not help and the patient's condition began to deteriorate – his hair fell out and his body was covered with sores. A sore even appeared on his tongue. The patient, who was only 20 years old, was completely exhausted, because he could not eat anything. Convinced that even the most modern medication did not help, the young man's relatives took him to Mecca for a small pilgrimage (Umrah). Four days after Talal began procedures with Zamzam water, the sores on his body began to heal and after another four days, they disappeared altogether, including the sore on his tongue. He returned home and his condition gradually started to improve (Huseynov). This may be considered a case in which, contrary to the expectations of the doctors, when this man recovered fully he and his relatives considers it as a miracle of Zamzam water. This water, which was sent down to people by God, brings only happiness to those who taste it.

Dry weather makes Zamzam more saline through evaporation, which is good for the human body. Zamzam water has always maintained the same taste ever since it came into existence. Its potability has always been universally recognised. Pilgrims from all over the world visit Ka'ba every year for Hajj and Umrah and have never complained about it.

Modern science has proved that the water of Zamzam is different from all other types of water in the world. Pilgrims to Makkah collect Zamzam water it in bottles to bring for relatives and friends back home who are ill.

Conclusion

On the basis of these experiences, I conclude that people have faith in the supernatural powers of Zamzam water and drink it for their wishes to be fulfilled. Likewise, it may be emphasised that Indian rivers have also acquired a distinct sacredness and this power is manifested in their free flowing nature, which attract a large number of pilgrims coming from distant parts to pay their homage.

Islam treats water as a life giving substance of nature like other religions. It has hidden qualities which makes it sacred in the eyes of believers of different religions.

References

Griffin (2007), S O, E Regnier, P M Griffin, V Huntley, *Effectiveness of fluoride in preventing caries in adults*– Journal of Dental Research 86 (5): 410–15.

Haleem (1989), A.M., "Water in the Quran," *Islamic Quarterly*, 33 (1), 34-50.

Hawting (1980), G. R., "The Disappearance and Rediscovery of Zamzam and the 'Well of the Ka'ba'". *Bulletin of the School of Oriental and African Studies*, University of London 43(1): 44–54.

Huseynov, A. *Zamzam – life-giving water*, *Region Plus*. Azerbaijan. info@regionplus.az

Hussain (2010), T. & M. Ahmed, *Zamzam Water: Scientific research findings*. Al-Islaah publications.

Zam Zam Studies and Research Centre (2005). Cited from The Saudi Geological Survey.

3

Water in Christianity

Fr Subhash Anand

Editor's Introduction: The author describes how sprinkling of holy water is said to remove the sins and renew and rejuvenate a person in the Christian tradition.

This happens because water has the capacity to transfer spiritual charges to the receiver. Water has become part of symbolic discourse.

Jesus was baptised by John in River Jordan. This River is considered Holy by Christians.

To understand the role of water and its symbolic function within Christian life we need to go back to the Bible, as that is the primary source of Christian symbolism. We also need to keep in mind that symbols are ambivalent. They portray the pleasant and the unpleasant, the acceptable and the unacceptable, the positive and the negative aspects of our life. They are also polyvalent. One and the same symbol can evoke more than one positive or negative element in our experience.

The Symbolism of Water in the Christian Tradition

Water as Life

We need water to quench our thirst or we will be dehydrated and even die. The Lord cares for his devotee and so he leads him to a place where he can find water (Ps 23.2). The Lord even draws water from the rocks to preserve his devotees (Ps 78.20). On the other hand, a person who feels abandoned by God finds himself in a dry and thirsty land where there is no water (Ps 63.1). Our animals and plants need water. A river is a good source of water, and so it evokes sentiments of security and joy. The person who is blessed by the Lord is like a tree planted by the river side: he will always thrive (Ps 1.3).

Our longing for water becomes a symbol of our longing for God. "As a hart longs for flowing streams, so longs my soul for thee, O God. My soul thirsts for God, for the living God. When shall I come and behold the face of God?" (Ps 42.1-2). A person who is far away from God is like parched earth. "O God, thou art my God, I seek thee, my soul thirsts for thee; my flesh faints for thee, as in a dry and weary land where no water is" (63.1). "I stretch out my hands to thee; my soul thirsts for thee like a parched land" (143.6). The Lord himself invites us to him: "Ho, every one who thirsts, come to the waters" (Is 55.1).

Since water is so much related to life it becomes significant even in the Biblical story of creation. "In the beginning God created the heavens and the earth. The earth was without form and void, and darkness was upon the face of the deep; and the Spirit of God was moving over the face of the waters" (Gen 1.1-2). Here the waters are evocative of the primordial womb, being fecundated by the divine breath of life. The first human couple is placed in a beautiful garden that is watered by a river–the primordial river, which on leaving the garden becomes the four great rivers known to the writer (2.10-14). These rivers flow in different four directions. The garden is symbolic of the beautiful earth. It is watered in all directions, and so it is full of life.

Water as Death

Water can be the bearer of death. Floods caused by heavy rains, overflowing rivers, tsunami waves, bursting of dams, etc., are a common phenomenon even today, and there is a great loss of life and property. In the book of Genesis we have an account of a great deluge. The whole earth gets submerged into water and all animals and even birds as die, as even the trees are covered with water. The deluge is interpreted as divine punishment for sin (6.7-7.24). On the other hand, God protects the righteous from the floods: "Therefore let every one who is godly offer prayer to thee; at a time of distress, in the rush of great waters, they shall not reach him" (Ps 32.6). The destructive effect of a great quantity of water suddenly coming rushing on us itself becomes a symbol of an invading army (Is 59.19; Jer 46.8). Even when there are no floods people can die due to drowning. A person abandoned by God feels that he is drowning: "Out of the depths I cry to thee, O Lord!" (Ps 130.1)

In some way the Old Testament anticipates what is actually happening now: our lakes and rivers are getting polluted. The water is becoming unfit not only for humans but even for aquatic creatures. In the book of Exodus, when the Pharoah (King of Egypt) refuses to allow the Israelites to leave Egypt, Moses turns the water of the Nile into 'blood' so that the Egyptians could not drink it and all the fish in it died (7.17-21). The most memorable experience of water as death, however, is when the Egyptians pursue the Jews who are fleeing away from Egypt. All the Egyptians with their horses and chariots are drowned in the sea (Ex 14.23-28).

Water as Purifier

We use water to wash ourselves and to wash our clothes and the many other objects that are part of our day-to-day life. This explains why in almost all ancient cultures water is symbolic of purification even within the religious sphere. People flock to a holy river because they believe that a bath in its waters will purify them from their sins. Before Moses consecrates Aaron and his sons as priests, he washed them clear (Ex 29.4). A devotee, deeply conscious of his sinfulness, prays that the Lord may wash him clean (Ps 51.2, 7). Through the prophet Ezekiel, God promises to purify his wayward people by sprinkling clean water over them: "I will sprinkle clean water upon you, and you shall be clean from all your uncleannesses, and from all your idols I will cleanse you. A new heart I will give you, and a new spirit I will put within you; and I will take out of your flesh the heart of stone and give you a heart of flesh. And I will put my spirit within you, and cause you to walk in my statutes and be careful to observe my ordinances" (36.25-27). This purification with water is not only at the physical level but also at the psychic or spiritual level. God says, "I will put new spirit within you" in the process of sprinkling clean water. The 'clean' water carries certain spiritual significance by becoming part of symbolic discourse. The faith of the person makes him see it as indicative of God's action in his life.

The Symbolism of Water in the New Testament

Water as Life

Jesus tells Nicodemus, who comes to him at night for fear of the enemies of Jesus, that he needs to be reborn of water and the Spirit (Jn 3.5). John's Gospel, which in different ways takes us back to the account of creation in the Old Testament, is referring to the hovering of the divine Spirit over the water mentioned there. Jesus is inviting Nicodemus to be become a new creation. Jesus himself offers the living water to those who come to him. On drinking this water we will not thirst again, for we will attain eternal life (Jn 4.13-14). This life-giving water flows from the heart of the Risen Jesus, and is nothing else but the gift of the Holy Spirit, the divine breath, that is the source all life (Jn 7.37-39). Water offered by Jesus is symbolic of his invitation to enter into communion with him. It is this personal communion that changes the inner psyche of the person thus Nicodemus becomes a new creation and enters eternal life.

Just as the Bible begins with a beautiful garden so to it ends with a beautiful garden, where a river flows and its water makes trees bear fruit all through the year (Rev 22.1).

Water as Death

Water is also symbolic of death and destruction. An evil spirit is believed to cast its victim into fire and water to destroy him (Mk 9.22). Here we see very clearly the ambivalent character of two fundamental symbols: fire and water. We need fire and water, warmth and moisture for life. Here they are the source of death. We can drown in water and die. This explains why the disciples of Jesus become panicky when during a storm their boats are flooded with water (Lk 9.23). Drinking polluted water too can cause death (Rev 8.11)

Water as Purifier

John is preaching in the desert, inviting people to repent. He baptizes them in the river Jordan for the remission of their sins and (Mk 1.4). The purifying symbolism of water makes it an instrument of healing: it purifies us from our sickness (Jn 5.2-7). Here again we see that water purifies one from sickness.

This happens because communion with Jesus brings us peace and joy and thus removes the psychic disturbances which are the cause of physical ailments.

Before his arrest, Jesus washes the feet of his disciples so that they could be clean and have a share in his life (Jn 13.5-8). In the early Christian community water was used in baptism because of his rich symbolism (Acts 8.36): Death to the old self through freedom from sin and living in Jesus.

The Symbolism of Water in Christian Worship

Water in Easter Liturgy

The liturgy (public prayer service) that marks Easter Vigil (the eve of Easter Sunday – the most important feast for Christians, celebrating the resurrection of Jesus from the dead) has a very special blessing of water. The prayer of blessing is even more solemn if there are some people to be baptised or if the water in the baptismal font is to be blessed. During the blessing, the Easter Candle, already blessed in the earlier part of the service, is dipped into the water. This candle is symbolic of the Risen Jesus, and the gesture reminds us of Jesus stepping into the Jordan to be baptised at the hands of John. The Christian tradition states that by coming into contact with Jesus not only the water of Jordan but all water has become holy, and a renewed source of life. Historians of religions see in the gesture of dipping the candle into the water a phallic and fecundating evocation. The baptismal font itself is seen as a womb, for through baptism people are born into the Church.

Water in Baptism

In some Christian denominations people are baptized by being immersed in water. For this the people may even go to a river that has clean water, or use a tank especially constructed for this rite. The person enters the tank from one end and emerges from the other. This is evocative of being buried with Jesus and rising with him. Thus the water in baptism has both a death and life symbolism: death to the old self and new life in Jesus. Christians also believe that in the first human couple Adam and Eve all humans have incurred original sin. We are washed clean in baptism.

Water in Blessings

In the day-to-day life of Catholic Christians, there are many occasions when we approach a priest for a blessing: when we move into a new house or buy a new vehicle; when our harvest is ready or an animal or a woman has given birth; when some person is sick or dead, etc. In all these blessing it is customary to sprinkle water on the person or object to be blessed. Even the corpse is blessed and sprinkled with blessed water before burial, and so is the grave to be used for the burial. This is, so to say, the final purification, and the preparation for a new birth into life eternal. The grave then is symbolic of the womb. This water itself is prepared with a special prayer of blessing and has salt dissolved in it. As intimately linked with life and purity, water is auspicious. The presence of the salt brings in an added symbolism: we pray that God may preserve the person or object blessed from all corruption. In some churches, at the beginning of Sunday worship water and salt are blessed and mixed together. This water is then sprinkled on the people assembled for prayer, to remind them of their baptism.

Conclusion

Water is symbolic of life in its totality. It is indicative of divine grace that gives us life, and helps us to live in a healthy manner, free from sin and sickness. In early ages we collected water from a river. Hence in primal consciousness a river is linked with life in all its aspects.

PART 2: IMPACT OF DAMS ON SUBTLE QUALITIES OF WATER

Introduction

Science and rationality exist in the realm of the mind, conscious or logic. Spirituality exists at the level of heart, unconscious or intuition. It is difficult to find scientific evidences of spiritual qualities of water because the basic concepts or categories with which spirituality works are different than those used by science.

Nevertheless, recent advances in science have enabled mankind to catch a glimpse of the spiritual qualities of water. There is an interface physical and between the spiritual qualities of water. The physical qualities of water are influenced and changed in a subtle way by its spiritual content.

The two chapters in this section provide some indication of the changes that take place in water due to spiritual charges.

The first article by Jhunjhunwala provides photographic evidence. Molecules of water do not exist in a chaotic state. They make intricate six-cornered clusters. The design of these clusters is influenced by thoughts of people around it. It can be posited that similar changes would take place by spiritual charging of water. This article provides an overview of the method and some evidence regarding impact of dams.

The second article by Paolini provides direct evidence of the impact of Tehri Dam on the subtle qualities of water.

The third article by Radin and others provides results of an experiment that indicates that molecular clusters of water are influenced by human thoughts.

4

Impact of Thoughts on Water Crystals

Bharat Jhunjhunwala

Editor's Introduction: In the previous section we have seen that all religions believe that water has the capacity to absorb and transfer spiritual charges. However, as of now there is very little scientific understanding of spirituality. More so, there is little understanding how water may become the medium of spiritual activities.

In this article the author explains how the molecular clusters in water may be the medium of its spiritiual powers. Japanese Scientist Masaru Emoto has developed a technique by which one can take photographs of molecular clusters of water. Molecules of water do not exist in a chaotic state. They make six-cornered clusters having different designs. The design of these clusters is influenced by the thoughts of people around it. It is possible that the spiritual charges are absorbed by water by it making special type of clusters.

The photographs of water crystals made from springs and free-flowing rivers are beautiful while those taken from water impounded in dams are not so. This points to dams negatively impacting the spiritual powers of water.

Japanese scientist Masaru Emoto has taken photographs of crystals made from water taken from different sources. Photographs of springs and clean rivers show beautiful forms while those of reservoirs and polluted rivers show disparate forms. This finding supports the view that hydropower projects will negatively impact the subtle quality of water.

In this chapter I first show that molecules of water form six-cornered clusters. Then I show that the shape of these clusters is influenced by thoughts of human beings. In the next section I show that the beautiful clusters in free-flowing rivers are destroyed in reservoirs indicating a negative impact. Lastly, I discuss the criticism of Emoto's method and provide my explanations.

Six-cornered Molecular Clusters

Kenneth Libberecht of Caltech has explained the molecular structure of water:[1] "The water molecules in an ice crystal form a hexagonal lattice, as shown below. Each red ball represents an oxygen atom, while the grey sticks represent hydrogen atoms as shown in Fig. 1. There are two hydrogens for each oxygen, so the chemical formula is H_2O.

"The most basic form of a snow crystal is a hexagonal prism. This structure occurs because certain surfaces of the crystal, the *facet surfaces*, accumulate material very slowly.

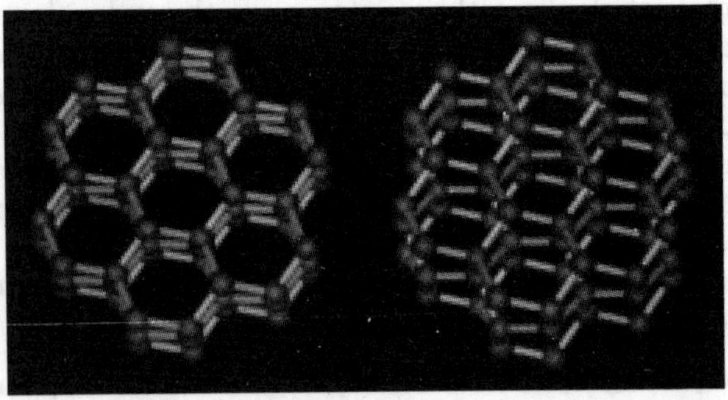

Fig. 1: Water molecules in an ice crystal.

"A hexagonal prism includes two hexagonal "basal" faces and six rectangular "prism" faces, as shown in the figure. A hexagonal prism can be *plate-like* or *columnar*, depending on which facet surfaces grow most quickly as shown in Fig. 2.

"When snow crystals are very small, they are mostly in the form of simple hexagonal prisms. But as they grow, branches sprout from the corners to make more complex shapes.

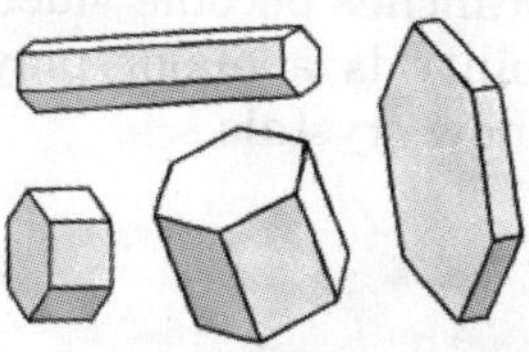

Fig. 2: Simple hexagonal prisms in small water crystals.

"By growing snow crystals in the laboratory under controlled conditions, one finds that their shapes depend on the temperature and humidity. This behaviour is summarised in the diagram at Fig. 3, which gives the crystal shapes under different conditions.

"The morphology diagram tells us a great deal about what kinds of snow crystals form under what conditions. For example, we see that thin plates and stars grow around -2°C, while columns and slender needles appear near -5°C. Plates and stars again form near -15°C, and a combination of plates and columns are made around -30°C. Furthermore, we see that snow crystals tend to form simpler shapes when the humidity is low, while more complex shapes are formed at higher humidities. The most extreme shapes — long needles are formed around -5°C and large, thin plates are formed around -15°C and when the humidity is especially high.

"Complexity of the water crystals arises because water molecules travel through the air or water to condense onto a growing crystal. Water molecules cluster into intricate designs. Consider a flat ice surface that is growing in the air or water. If a small bump happens to appear on the surface, then the bump sticks out a bit farther than the rest of the crystal. This means water molecules from afar can reach the bump a bit quicker than they can reach the rest of the crystal, because they don't have to diffuse quite as far.

"With more water molecules reaching the bump, the bump grows faster. In a short time, the bump sticks out even farther than it did before, and so it grows even faster. This is called branching instability – small bumps develop into large branches, and bumps on the branches become sidebranches. Complexity is born. This instability is a major player in producing the complex shapes of snow crystals.

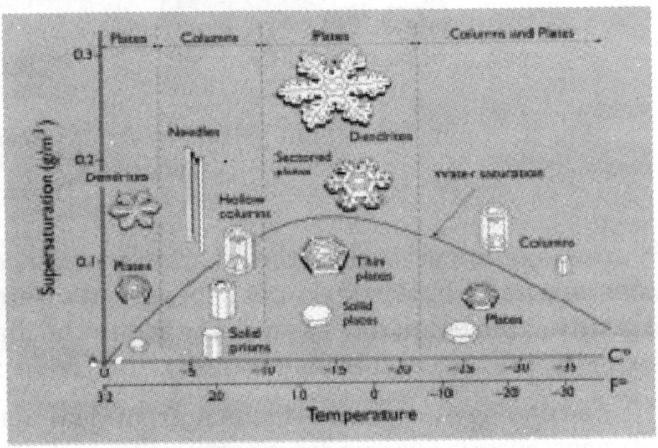

Fig. 3: Impact of conditions on shape of water crystals.

"When the branching instability applies itself over and over again to a growing snow crystal, the result is called an ice dendrite. The word dendrite means "tree-like."

"The growth of snow crystals depends on a balance between faceting and branching. Faceting tends to make simple flat surfaces, while branching tends to make more complex structures. The interplay between faceting and branching is a delicate one, depending strongly on things like temperature and humidity. This means snow crystals can grow in many different ways, resulting in the great diversity we see in snow crystal forms."

The water in snow flakes forms crystals in this way. These have been photographed by Libbrecht. Some photographs are reproduced below:

The most important finding here is that molecules of water invariably make six-cornered clusters.

Fig. 4: Photographs of water crystals from snow flakes taken by Libbrecht.

This six-cornered structure of snow flakes has been long known. Wilson Alwyn Bentley was the first photographer of snow flakes. He died in 1931. He photograped the snow flakes before they could melt. Some photographs taken by him are reproduced below:

Fig. 5: Photographs of snow flakes by Bentley.

The following points emerge from this study of snow flakes:

1. Water molecules make hexagonal structures.
2. These structures become more complex as crystals grow.
3. Complexity and shape of the crystals depends upon various factors such as temperature and humidity in air.

Impact of Thoughts on Water Crystals

The variation in shapes of the six-cornered water crystals do not adequately explain the designs in a crystal. Thoughts or 'vibrations' have a huge impact on the design. Emoto holds that words have different vibrations which influence the structure of the molecular clusters. In an interview Emoto explains:[2]

> REIKO: Does that mean that every word has its own signature vibration or cluster that is unique to itself?
>
> DR. EMOTO: Yes. During our evolution, we learned what sounds were dangerous, what sounds were soothing and safe, and what sounds were pleasurable, and so on. We slowly learned about various vibrations of the laws of nature. We learned this through instinct and through experience. We accumulated this information over time. We started out with some simple sounds like "a" or "u" or "e," which evolved into more complex sounds like "love." And these positive words create "natural" crystalline structures—which are all based on the hexagon.

Emoto says that persons and words have different frequencies of vibrations. These vibrations affect the molecular clusters of water.

> Most of the objects found in nature emit stable frequencies. Each sparrow sounds basically the same (though sparrows themselves might recognise slight differences), and the sounds made by dogs or cats do not have a great deal of variation. By contrast the human being is able to make full use of the *do, re, me, fa, so, la, ti* scale to create beautiful melodies.[3]

These beautiful melodies seem to create correspondingly beautiful molecular structures in water. These are the subtle qualities of water. The impact is not restricted to sounds made by human beings. The impact works both ways. Thoughts influence molecular clusters and molecular clusters influence thoughts.

From Emoto's work we are provided with evidence that human vibrational energy, thoughts, words, ideas and music, affect the molecular structure of water. Remember that water comprises over seventy percent of a mature human body and covers the same amount of our planet.

The body is like a sponge and is composed of trillions of cells that hold liquid. The water in these cells forms different structures and influences our life. Thus the quality of our life is directly connected to the quality of our water.

The vibrations of the environment will change the molecular shape of water. In this sense water not only has the ability to visually reflect the environment but it also molecularly reflects the environment.

Emoto has been visually documenting these molecular changes in water by means of his photographic techniques. He freezes droplets of water and then examines them under a dark field microscope that has photographic capabilities.

After seeing water react to different environmental conditions, pollution and music, Emoto and colleagues decided to see how thoughts and words affected the formation of untreated distilled water crystals using words typed onto paper by a word processor and taped on glass bottles overnight.

The same procedure was performed using the names of deceased persons. The waters were then frozen and photographed.

With the recent popularity in music therapy, Emoto decided to see what effects music has on the structure of water. He placed distilled water between two speakers for several hours and then photographed the crystals that formed after the water was frozen.

These pictures indicate that both thought and sound have a deep influence on the molecular structure of water.

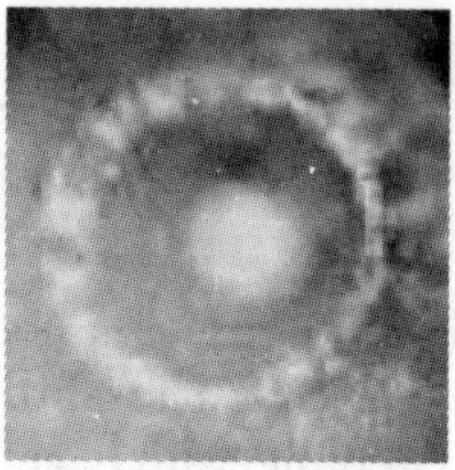

Untreated Distilled Water.

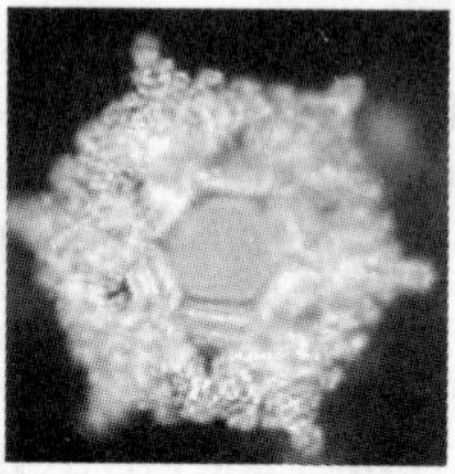

Love and Appreciation.

Thank you.

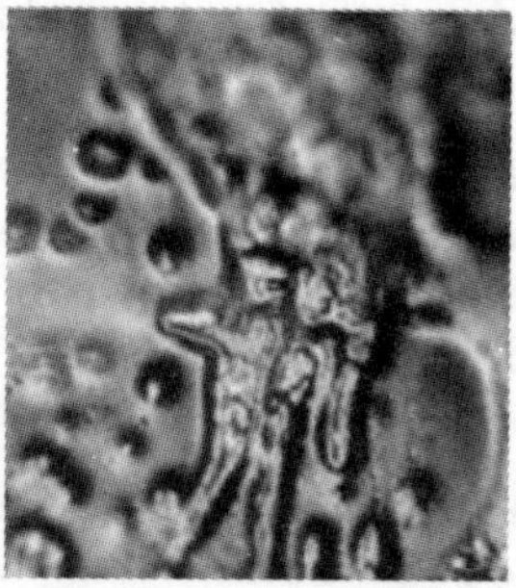

You make me sick, I will kill you.

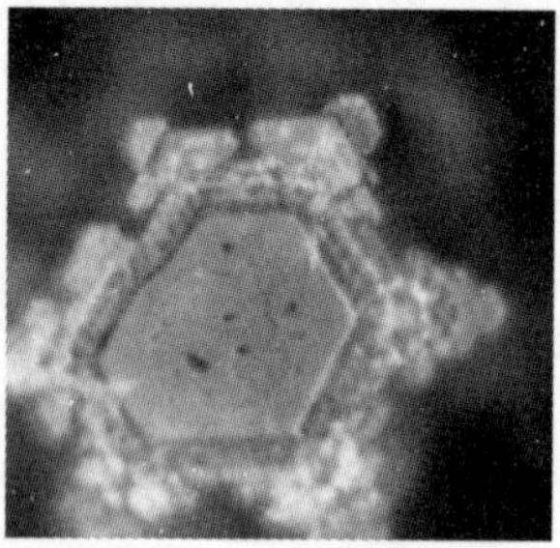

Mother Teresa.

Bach's 'Air for the G string'.

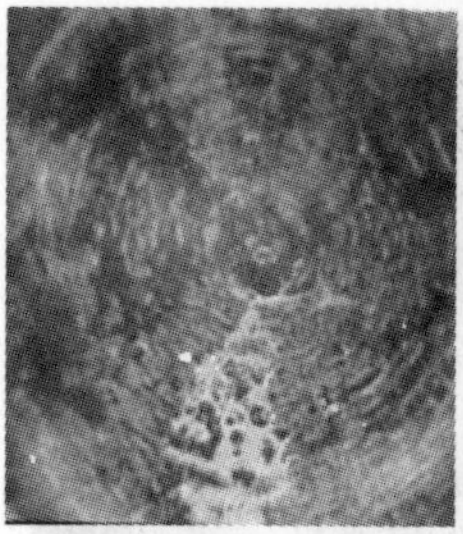

Heavy Metal Music.

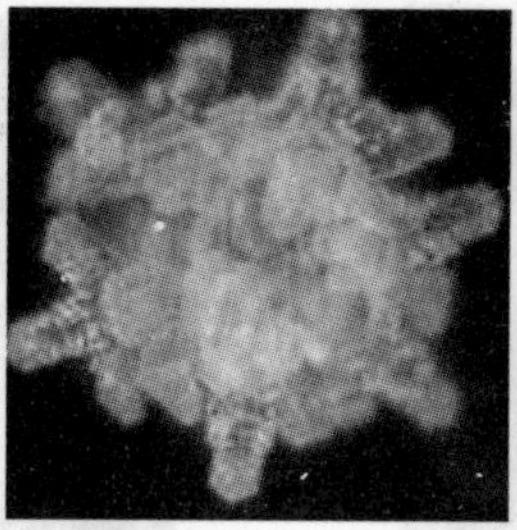

Tibet Sutra.

Kawachi Fold Dance.

Impact of Dams on Molecular Structure of Water

Emoto has taken photographs that show the impact of damming a river on its molecular structure. Images of free flowing water are given below.

The structure is altogether different in dams and polluted rivers.

These photographs show that the beautiful molecular structure of water is destroyed when water is held up in reservoirs. I assume that impact of hydropower turbines would be similar. It would be good if such photographs could be taken upstream and downstream of a hydropower project. However, I have not seen this yet.

Sanbu-ichi Yusui Spring Water.

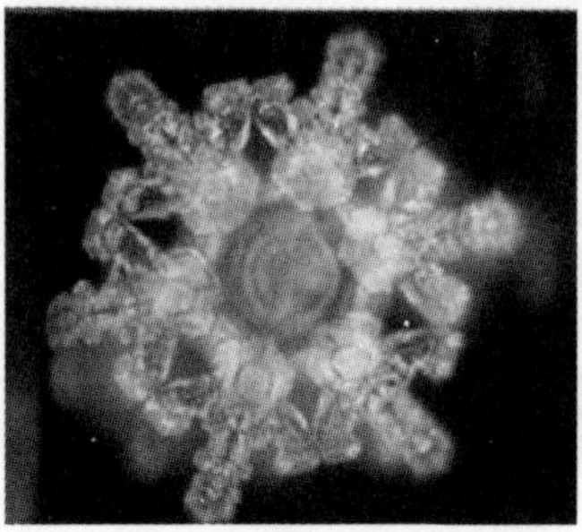

Japan Shimanto River, referred to as the last clean stream in Japan.

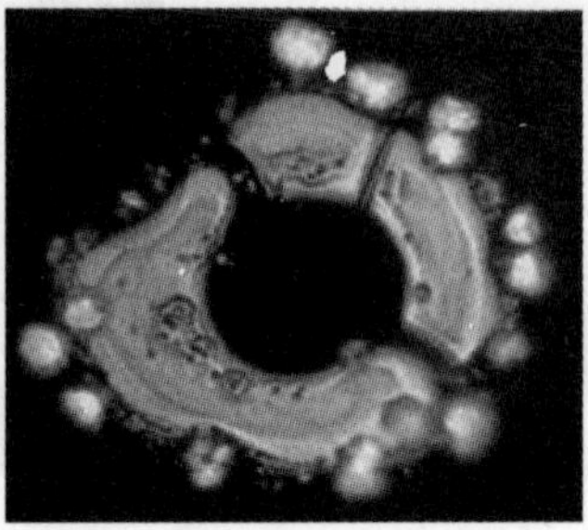

Biwako Lake, the largest lake at the center of Japan and the water pool of the Kinki Region. Pollution is getting worse.

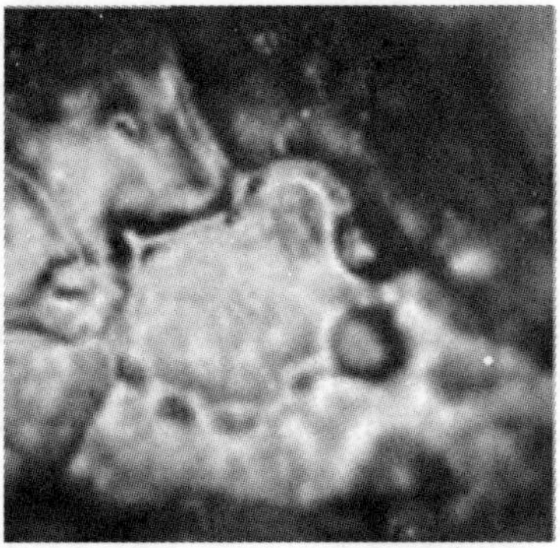

Fujiwara Dam, before offering a prayer.

It can be contended that the above photographs are not conclusive in the sense that the impact of change in molecular structure on the quality of water felt by human beings is not established. In this regard I present findings of a survey of pilgrims that I had got done. The objective of the study was to assess whether and to what extent the construction of Tehri dam has affected the spiritual power of River Ganga. A questionnaire was made to ask respondents about impact of dam on various aspects of worship. Naresh Chandra Puri, Diploma Pharma, resident of Village Lakshmoli, Dt Tehri, went to these pilgrim centres in December 2007-January 2008 and sought responses from the pilgrims coming to take bath or to collect water for taking home.

No sampling was done. Assistance of any local person was not sought lest the mediator influence the results. The enumerator asked people he met spontaneously about the impacts. The responses given by the pilgrims are tabulated below.

Table 1: Question: What is the impact of Tehri Dam on flow of waters of Ganga? (Per cent of Pilgrims Responding in Affirmative)

Sl. No.	*Type of Pilgrim*	*Was Better Previously*	*Is Better Now*[4]	*No Difference*	*Net, Per cent Reporting Previously Better*
1.	Dev Prayag, n=41	88	12	-	76
2.	Rishikesh, n=87	47	15	38	32
3.	Haridwar, n=78	38	26	36	12

* At Dev Prayag the question was phrased as follows: What will be the change in spiritual power of water after extraction of electricity?

Conclusion from the above Table is that there is decline in quality of water due to construction of Tehri Dam. However, the decline is less as we move away from Tehri. This is likely because the water gets a chance to regenerate some qualities.

Another question was asked to cross-check the responses.

Table 2: Question: What will be the benefit to the nation if Tehri Dam is removed and pilgrims are able to take bath in free-flowing waters of the Ganga at the Confluence? (Per cent of Pilgrims Responding in Affirmative)

Sl. No.	*Type of Pilgrim*	*Small Increase*	*Big Increase*	*Small Decrease*	*Big Decrease*	*Net, Per cent Increase Reporting Increase*
1.	Dev Prayag, n=34	44	47	6	3	82
2.	Rishikesh, n=78	36	45	11	8	62
3.	Haridwar, n=81	38	32	18	12	40

Conclusion again is that there is a positive impact on pilgrims if Tehri Dam is removed but the intensity of impact declines as we move downstream.

This study indicates that dams indeed lead to a negative impact on the subtle quality of water. This could be due to the change in molecular structure of water that has been shown to take place.

It can still be said that this evidence is not 'conclusive' since the study was not done by a credible institution. In this regard, Indian law provides for the 'Precautionary Principle.' The idea is that in case there is unproved suspicion of some negative environmental impact, then it is better to err on the side of caution. In keeping with this principle, making of dams on rivers should be reconsidered in order to prevent a possible irreversible loss.

Emoto's Method and its Criticism

Emoto's method of taking pictures has been much criticised. It is useful to outline his method before dwelling into its criticism.

> Emoto's method of obtaining photographs of crystals involves a relatively simple and inexpensive process. 0.5 ml samples of liquid water from a specific sample are placed on 100 petri dishes that are then frozen and stored at a temperature of -25°C for three hours in a freezer. A sample is removed from the freezer for observation under a microscope with a camera in a room with a constant temperature of -5°C. As the microscope's light melts the top of the sample, crystals are observed and photos are taken.[5]

Emoto's method has been criticised by many. Setchfield is perhaps the most vocal critic. His main point of criticism is that the procedure has no means of ensuring that experimenter's bias is prevented or minimised. The methodology does not ensure that the obtained results are not selected consciously or subconsciously by the photographer. A double blind procedure in which a photographer would not know what water sample he or she was photographing would make the claim considerably more credible, he says.

Fortunately, precisely such a double blind procedure has since been conducted. Summary of the results are as follows:

> An experiment tested the hypothesis that water exposed to distant intentions affects the aesthetic rating of ice crystals formed from that water.
>
> Over three days, 1,900 people in Austria and Germany focused their intentions towards water samples located inside an electromagnetically shielded room in California. Water samples located near the target water, but unknown to the people providing intentions, acted as "proximal" controls. Other samples located outside the shielded room acted as distant controls. Ice drops formed from samples of water in the different treatment conditions were photographed by a technician, each image was assessed for aesthetic beauty by over 2,500 independent judges, and the resulting data were analysed, all by individuals blind with respect to the underlying treatment conditions.
>
> Results suggested that crystal images in the intentionally treated condition were rated as aesthetically more beautiful than proximal control crystals (p ¼ 0.03, one-tailed).[6]

Full report of the study is placed at Chapter 6. This study confirms that thoughts influence the structure of water. Bias could influence the results but the impact of thoughts on water crystals stands out even if the specific results are biased.

Another attempt to replicate Emoto's research was made by Advanced Placement Psychology class at Durango High School. Emoto holds that different messages stuck to the water container influence the molecular clusters of the contained water. The students stuck different messages and then looked at the structure of molecular clusters. They concluded:

> We did not find sufficient evidence to refute or accept Emoto's hypothesis that thought influences water crystal formation. We noticed one interesting similarity between two separate groups of water samples: Similar crystals formed on the same message, "I despise you." But, for the most part, the crystal formations in each water sample resembled each other, regardless of the messages attached

to them. We concluded that in order to make a significant finding, further research would have to be done.[7]

A close reading of this conclusion actually supports Emoto's findings. The students did find "similarity between two separate groups of water samples: Similar crystals formed on the same message, 'I despise you'." Indeed, they did not find differences in other samples. But this could be due to shortcomings in the experimental processes. Absence of this finding does not negate the similarity observed in the crystals made from the message 'I despise you.'

Other criticism made by Setchfield and my under-standing of these are given below.

Criticism: Ice crystal structural formation is dependent on numerous environmental factors, the most important of which are temperature and humidity. While Emoto minimised some possible sources of error by conducting his studies in the same room with the same sample sizes, the same freezer and same microscope each time, other possible sources of error were not addressed. For example the Petri dishes were not sealed to prevent contamination or disturbance by the operator or environment. A simple thing such as the photographer's breath while using the microscope could affect the warming rate of the frozen sample and temperature of crystal formation, thus affecting the structure of the resultant crystal.

Response: Indeed the results could be influenced by such other factors. That, however, is only a possibility. The positive evidence cannot be discarded on this ground.

Criticism: Dr. Emoto's procedures indicate that his samples are frozen at -25°C, and his ice crystals are formed at -5°C. These temperatures should produce mostly column crystals rather than plate crystals, yet not one of Emoto's published photos show a column crystal (see Fig. 3 above). This makes Dr. Emoto's data suspect (as they appear to conflict with the findings of well-respected researcher) and indicates the possibility that Emoto excluded non-supportive data from his publications.

Response: Fig. 3 above shows that plates are formed at 0 to -4 degrees. However, I am not competent to comment on this.

Criticism: While Emoto has published his claim in one peer reviewed journal, it has neither been substantiated nor disproved by research scientists.

Response: The findings cannot be dismissed on this ground. The disinterest of the scientific community may be due to their bias against such views.

Criticism: Emoto may be driven by his commercial interests. Indigo water that is "highly charged hexagonally structured concentrate," and supposedly creates "structured water" that is "is more easily assimilated at the cellular level" is being sold by him for $35 for an eight-ounce bottle.

Response: Many Pharmaceutical Corporations undertake research with commercial interests. That is never decried. We should look at the science and not get distracted by the existence of such motives.

My conclusion is that while Emoto's findings need more substantiation, there is nothing available that negates his findings. All the available evidence is either non-confirmatory or positive. Libbrecht's and Bentley's photos of snow flakes are very similar to the photographs taken by Emoto. This supports the key statement that water molecules form six-cornered clusters and that the design of these is changed by certain influences. Only the precise nature of these influences needs more substantiation. This establishes the basic soundness of Emoto's method. Hence we should accept his findings until disproven otherwise.

References

1. Large parts of this section in the paper have been reproduced from Libbrecht's paper with due acknowledgement: Libbrecht, Kenneth G., *Snow Flake Primer*, Caltech.
2. Wellness Goods, Miraculous Messages From Water, http: //www.life-enthusiast.com/twilight/research_emoto.htm
3. Emoto, Masaru, *Hidden Messages in Water,* p. 51.

4. The term 'flow' focuses more on quantity of water than quality. There is more quantity of flow in winters and summers due to construction of Tehri Dam. This may be a positive aspect that was noted by Pilgrims. This positive impact, in my assessment, should not be taken as a positive impact on the spiritual quality of the water.
5. Kristopher Setchfield, Is Masaru Emoto for Real???, http://www.is-masaru-emoto-for-real.com/
6. Radin, Dean, Nancy Lund, Masaru Emoto And Takashige Kizu, Effects of Distant Intention on Water Crystal Formation: A Triple-Blind Replication, Journal of Scientific Exploration, Vol. 22, No. 4, pp. 481–493, 2008.
7. Damian Nash, an experiment attempting to replicate Dr. Emoto's work..., May 26, 2005, http://www.amazon.com/review/R3BQBA6BI653IJ.

5

Impact of Tehri Dam on Energy in Water

Tiziano Paolini

Editor's Introduction: Swiss Laboratory Aqua Viva has taken photographs of water crystals made from water of the Ganga upstream and downstream of Tehri Dam on the asking of Ganga Aavahan, Uttarkashi. Analysis of these photos shows that water has higher energy before entering the Tehri Dam and less energy at exit. This indicates that energy gets lost in generation of hydropower.

The first two pictures given below are taken upstream Tehri Dam. These show brightness and energy. The third picture is taken downstream Tehri Dam. Lack of energy and contraction is seen. Comparison of the first two with the third picture indicates that energy of water is lost in the Tehri Dam.

The pictures of the water crystal are taken examining a droplet of frozen water under the microscope. The crystal appears on the tip of the droplet and one must have the ability to choose the right spot and moment to take the picture. The dimensions we work with are in thousandth of a millimeter therefore precision is a very important factor.

The shape of the crystal is given by the union of molecules through the hydrogen bond. Molecules receive an 'order from above' to arrange themselves in a certain way and this 'order' is a high-frequency wave which no scientific device can perceive while water can.

If we expose grains of sand lying on a metal plate to the vibration of a violin bow we will observe the grains arranging themselves in a specific geometrical form.

If by analogy we substitute the grains of sand with the molecules, and the sound vibration with a 'vibration from above' (a super-sensory/celestial vibration), we will have the pictures you are looking at. In fact water has the ability to make visible a dimension of reality which has been upto now unknown.

Our laboratory carries out these tests only when the time is deemed right and propitious, as we are aware that any negative subjective state of consciousness (such as agitation, stress, or excitement) may jeopardise the results.

On our side there was a bias when we began this work, as previous information had described the Ganges as one of the most polluted rivers in the planet.

It was therefore a surprise and a wonder that in nearly all these tests the water has manifested the shape of an hexagonal crystal that is normally found only in pure spring waters. The presence of this hexagonal shape means that this water maintains its original characteristics through contact with its source and that polluting substances or other negative energies do not contaminate it due to the great amount of worship and prayer of which it is recipient.

It is therefore evident that the positive thoughts sent to it by those who love the Ganges prevail over the negative chemical and physical characteristics that occur when the river is mistakenly seen as only matter.

Water is a vehicle of information capable of generating life; it is like a photo film that gets impressed with a flow of subtle electromagnetic waves well beyond those of visible light fields.

Water is also a substance capable of holding on to such information. It retains a memory which our tests make visible.

To ascertain that the water of the Ganges has maintained identical characteristics throughout its course has given me great satisfaction as it demonstrates that the forces of thought prevail over matter.

A chair is born out of a thought.

One may succeed in managing one's actions; managing one's words is already more difficult; managing thoughts still eludes us. But this is what we must learn if we want to avoid our Planet's destruction.

Water is life and as such must be protected globally and not be subjected to vested economic interests.

Being conscious of the benefic energy that runs through this River's waters one must conclude that an alteration of its flow would bring about negative consequences that we are presently too ignorant to estimate.

How and when is only known to the wise.

Hence the development of intuition is our task; prayer, meditation, and dreams are the means.

Picture 1: Gangotri: Hexagonal shaped crystal suggesting a tridimensional image with a luminous top.

Picture 2: Maneri Reservoir (upstream of Tehri Dam): Hexagonal shaped crystal growing an upward spiral along a vertical axis.

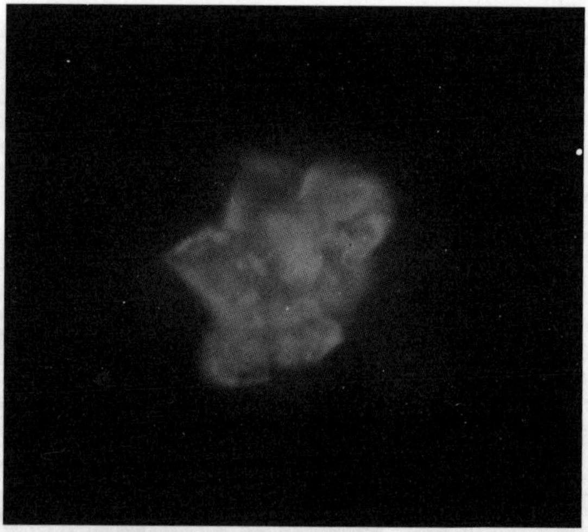

Picture 3: Bhagirathi Puram Zero Point + 0.5 kms (Downstream of Tehri Dam): Self-enclosed crystal showing a difficult expansion. Lack of energy in its unfolding.

6

Effect of Thoughts on Water[1]

Dean Radin, Gail Hayssen, Masaru Emoto and Takashige Kizu

Editor's Introduction: A major scepticism regarding Emoto's study is that the results are highly dependent upon the predisposition of the photographer. This study describes an experiment wherein photographs were taken blind. The results indicate a positive influence of thoughts on the patterns of water crystals.

The hypothesis that water "treated" with intention can affect ice crystals formed from that water was pilot tested under double-blind conditions. A group of approximately 2,000 people in Tokyo focused positive intentions toward water samples located inside an electromagnetically shielded room in California. That group was unaware of similar water samples set aside in a different location as controls. Ice crystals formed from both sets of water samples were blindly identified and photographed by an analyst, and the resulting images were blindly assessed for aesthetic appeal by 100 independent judges. Results indicated that crystals from the treated water were given higher scores for aesthetic appeal than those from the

1. *Explore,* September/October 2006, Vol. 2, No. 5.
Double-blind Test of The Effects of Distant Intention on Water Crystal Formation by Dean Radin, PhD, Gail Hayssen, Masaru Emoto, BA, and Takashige Kizu, BA.

control water ($P = 0.001$, one-tailed), lending support to the hypothesis.

Introduction

Experiments over the past four decades have investigated whether intention affects properties of water. This question is of interest to complementary and alternative medicine research, and especially for therapies involving intention, because the adult human body consists of approximately 70 per cent water. The question has been studied by comparing the effects of intentionally "treated" water versus untreated control water on the germination and growth of plants, including barley, wheat, rye, beans, cress, radishes, and lettuce. Other properties of water that have been examined include rate of cooling, molecular bonding as reflected by infrared spectra alterations, Raman spectroscopy, scattered laser light, and pH level. Although formal metaanalyses have not been performed on these studies, overall, the experiments provide evidence suggesting that various properties of water may be influenced by intention.

Interest in this topic has been rekindled recently by claims suggesting that intentionally influenced water can be detected by examining ice crystals formed from samples of that water. The specific claim is that positive intentions tend to produce symmetric, well-formed, aesthetically pleasing crystals, and negative intentions tend to produce asymmetric, poorly formed, unattractive crystals.

As of this study, no experimental replications of these claims have been published in peer-reviewed journals, so most critiques have assumed that the simplest explanation for the claimed results is either biased selection of crystal images or selective reporting of results or both. This paper reports a pilot investigation of the crystal formation hypothesis to test the claim under double-blind conditions.

Methods

Water Sample Preparation

In preparation for the experiment, the second author (G.H.)

purchased four plastic bottles of Fiji brand commercial bottled water (Los Angeles, CA). This brand was selected because, unlike many other bottled waters, after the Fiji label was removed, the plastic bottle contained no words, symbols, or other shapes embossed in the plastic. The first author (D.R.) randomly assigned the bottles with labels A through D (using a tossed die), and, with G.H., he selected two of those bottles to be the "treated" samples (randomly selected as A and B); the remaining two bottles were set aside as controls (C and D).

The treatment bottles were placed inside a double-steel-walled, electromagnetically shielded room (Series 81 Solid Cell; Lindgren/ETS, Cedar Park, TX) at the Institute of Noetic Sciences (IONS) in Petaluma, California. This room was used primarily as a convenient, limited-access location in which to place the bottles during the remote treatment period. The control bottles were placed in separate cardboard boxes and stored on a desk in a quiet location on another floor of the building that housed the shielded room. D.R. and G.H. did not inform the third or fourth authors (M.E. or T.K.) about the existence of the control bottles until after the treatment period was completed.

A digital photo of the two treatment bottles in the shielded chamber was e-mailed to M.E. and T.K. in Tokyo to be used as a visual aid for a group that would later be directing their intentions toward those bottles. The treatment and control bottles were maintained at approximately the same temperatures and were handled approximately the same length of time.

Treatment

On November 16, 2005, M.E. led a group of approximately 2,000 people in Tokyo in a prayer of gratitude directed toward the water in the IONS laboratory, some 5,000 miles away. M.E. showed the audience where the IONS laboratory was located in relationship to Tokyo by using a sequence of images from the Google Earth global mapping application (Mountain View, CA). Next, he showed the digital picture of the bottles inside the shielded chamber with the words of an intentional "prayer for water" overlaid on the photo. After explaining the photo and purpose of the experiment, M.E. led the group in speaking aloud the words of the prayer. This lasted approximately 5 minutes.

Table 1. Comparison of the Grand Mean Ratings for Aesthetic Appeal of the Treated vs Control Crystals by 100 Independent, Blind Judges

Aesthetic Appeal Ratings	*Treated*	*Control*
Mean	2.87	1.88
Variance	0.89	0.91
Observations	24	16
Pooled Variance	0.90	
Hypothesised mean difference	0	
df	38	
Ftest	3.27	
P value, one-tail	0.001	

Analysis

The day after the conference, D.R. and G.H. retrieved all four bottles and wrapped them in identical sheets of aluminum foil and bubble wrap, placed each bottle in a separate box labeled A through D as appropriate, and mailed the four boxes to M.E.'s laboratory. (The treatment and control bottles were in their separate locations for approximately 36 hours before being brought together again for packaging.) The bubble wrap and aluminum foil were used to provide a rudimentary shield against sudden shocks, ambient light, and electromagnetic fields that might have impinged on the bottles while en route to Japan. Each box was packaged separately to avoid the possibility that treated bottles might influence the control bottles through close proximity. After mailing the boxes, D.R. and G.H. informed M.E. and T.K. about the two control bottles, but they were not told the conditions of the four bottles to ensure that their analysis would be conducted blindly.

Upon receiving the four boxes, T.K. examined water samples from each bottle according to the following procedure:

1. For each bottle, approximately 0.5 mL of water was placed into each of 50 Petri dishes, and a lid was placed on each dish.

2. Each dish was then placed into a freezer maintained at —25 to —30°C for a minimum of 3 hours.
3. T.K. later removed the dishes from the freezer, and, in a walk-in refrigerator (maintained at — 5°C), he examined the apex of each resulting ice drop for a crystal using a stereo optical microscope. Previous experience with ice drops formed under these conditions indicated that the apex was the location where crystals were most likely to form. Crystals were defined as hexagonal shapes.
4. If a crystal was observed at the apex (not all ice drops formed discernable crystals), T.K. photographed it at either X100 or X200 magnification, depending on the size of the crystal.
5. All resulting photographs, from all four bottles, were then e-mailed to D.R.

Results

Analysis of Crystals

D.R. received a total of 40 photographs: 12 crystals were from bottle A, 12 from B, 7 from C, and 9 from D. Bottles A and B were the treated bottles, thus slightly more crystals were identified in the treated condition. To assess the aesthetic appeal of these 40 crystals, a group of 100 volunteers were recruited over the Internet to blindly and independently rate each crystal, one at a time, on a scale from zero to six, where zero meant "not beautiful" and six meant "very beautiful." (The ratings were collected via a Web site programmed in Perl/CGI by the first author.) Beautiful crystals were defined as symmetric, aesthetically pleasing shapes.

The order in which each image was presented to each judge (via a Web page) was independently randomised. In addition, the image titles (e.g., "2.jpg," "3.jpg") were assigned random numbers so the condition could not be inferred by examining the name of the crystal image and the condition (treated vs control) from which each crystal was formed was not indicated, and no feedback was provided at the end of the assessment exercise to prevent judges from learning which images corresponded to which conditions. The planned analysis was based on the first

100 judges to complete ratings of all 40 crystals, for a total of 4,000 contributed ratings.

Comparison of Ratings

Comparison of the mean ratings assigned to the images showed that the crystals from the treated water were rated significantly higher for aesthetic appeal than the crystals from the control water *(P = 0.001, one-tailed)*, as shown in Table 1 and Fig. 1.

Discussion

This pilot study was designed to test the most plausible conventional explanation for the crystal formation claim: the presence of subjective biases. To eliminate these biases, the person taking photos of the crystals (T.K.) and the aesthetic raters of those crystals were both blind to the treatment versus control conditions. The results were consistent with the hypothesis that water treated with pleasant intentions would result in more pleasing crystal shapes. If this effect was not due to obvious subjective biases, then what else might have accounted for the results?

Could the image assessment process have introduced a systematic rating bias? This question arises because each rater examined 40 crystal photos without prior training on the meaning of "aesthetically beautiful." If the sequence in which raters observed these images was in a fixed order, then this might well have introduced an artifact. However, this possibility was specifically prevented by presenting the images in a newly randomised order for each rater, thus averaging out potential sequential effects.

Another concern may be that the scales used to assess beauty, which were based on ordinal rather than interval measurements, violated the *t*-test assumption of an underlying normal distribution. This potential problem was ameliorated by using scores averaged across 100 raters, but parametric assumptions can be completely avoided by using a non-parametric statistic.[1] The method we used compared the observed mean difference in ratings between treated and control crystal images to the same mean difference determined after randomly

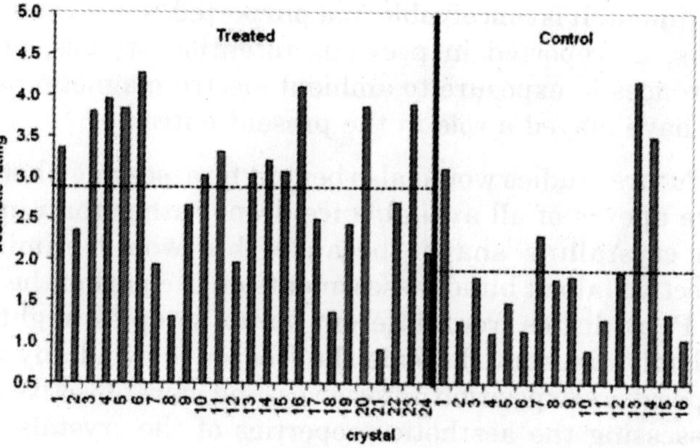

Fig. 1. Mean aesthetic ratings by 100 independent judges blind to the condition of each image, for 40 crystals blindly identified by an analyst. Grand means for the 24 treated and 16 control samples are identified with dashed lines.

reassigning which images belonged to the treated and control conditions. The original mean difference was larger than 999 out of 1,000 randomly reassigned differences, thus the associated P value was $P = 0.001$, virtually identical to the results of the £-test.

One might ask whether a new group of raters would replicate the initial results. This was tested by examining the results provided by 100 additional raters. The resulting t test was t (38 df) = 3.11, $P = 0.002$, confirming the original findings.

Perhaps the treated and control bottles were handled differently (say, with more fingerprints on the treated bottles), and the analyst (T.K.) detected those differences and was subsequently biased. Anticipating such artifacts, investigators D.R. and G.H. took care to ensure that the four bottles were physically handled the same way and then packaged identically. A related question is whether the treated and control environments might have differed in important ways. The treated bottles were in an electromagnetically shielded room during the treatment period, but the control bottles were not. In addition, that shielded room had been used exclusively for intention-related experiments for several years prior to this experiment. It is conceivable that purported "space conditioning" effects, as reported in previous intention studies, or simply differences in exposure to ambient electro-magnetic radiation, may have played a role in the present outcome.[2, 3]

Future studies would also benefit by assessing photographs of the apexes of all available ice drops rather than just those with crystalline shapes because this would eliminate all subjective (albeit blind) assessments on the part of the analyst. The Petri dishes from different water samples might also be randomly distributed inside the freezer to avoid any inherent temperature or position differences, and more objective methods of assessing the aesthetic properties of the crystals could be employed.[4, 5]

There are at least three unconventional alternatives that might explain the observed effects. One is that the intentional source was not the audience in Tokyo but rather D.R. and G.H. This possibility cannot be excluded, but, although these investigators were open to the hypothesis, neither held strong expectations about the experimental outcome. A second possibility is that the water was not altered at all, but rather the bottles were randomly assigned by D.R. and G.H. to the two conditions that would later result in a fortuitous differential effect or that T.K. fortuitously decided to take photos that would ultimately result in the observed outcome or both. Such anomalous assignment effects, formalised as "Decision Augmentation Theory,"[6] require the ability to unconsciously sense and act upon future possibilities, ie, a form of precognition. A third possibility is that the intentions of future observers (including readers of this article) *retroactively* influenced the water. Although this explanation may seem outrageous, there is experimental evidence suggesting that such time-reversed effects may exist.[7]

In conclusion, the present pilot results are consistent with a number of previous studies suggesting that intention may be able to influence the structure of water. Future replications should concentrate on eliminating all conceivable conventional artifacts, and protocols should be employed that can help discriminate among the various unconventional explanations.

References

1. Davison AC, Hinkley DV. *Bootstrap Methods and Their Application.* (Cambridge Series in Statistical and Probabilistic Mathematics, No 1). Cambridge, UK: Cambridge University Press; 1997.
2. Pyatnitsky LN, Fonkin VA. Human Consciousness influence on water *structure.*] *Sci Expl.* 1995; 9:89-105.
3. Radin DI, Taft R, Yount G. Possible effects of healing intention on cell cultures and truly random events. *J Altern Complement Med* 2004; 10:103-112.
4. Laviea T, Tractinsky N. Assessing dimensions of perceived visual aesthetics of web sites. *Int J Hum Comp Stud.* 2004; 60:269-298.
5. Gobster PH, Chenoweth RE. The dimensions of aesthetic preference: a quantitative analysis. *J Environ Manage.* 1989; 29:47-72.
6. May EC, Utts JM, Spottiswoode SJP. Decision augmentation theory: Towards a model of anomalous mental phenomena. *J Para-psych.* 1995; 59:195-220.
7. Braud W. Wellness implications of retroactive intentional influence: exploring an outrageous hypothesis. *Altern Ther Health Med* 2000; 6:3748.

PART 3: IMPACT OF DAMS ON PHYSICAL AND CHEMICAL QUALITIES OF WATER

Introduction

Supporters of hydropower say that generation of hydropower does not 'consume' water. Volume of water coming in and going out of a turbine is same. Therefore, we are able to generate electricity without consuming any fuel. Reports commissioned by the hydropower companies invariably conclude that there is no adverse impact of hydropower generation on the quality of water.

The first three papers are critiques of studies commissioned by the hydropower companies. These critiques show that actually there is a huge impact on the physio-chemical qualities of water. More importantly, these critiques are based on the same data that the authors of these reports used to conclude that there is no adverse impact. These papers show how the scientific establishment misrepresents data to deny the adverse impact.

This negative impact takes place in both types of hydropower projects—those involving inter-seasonal storage like Tehri; and those involving only intra-day storage like Vishnu Prayag on the Alaknanda. The impact of inter-seasonal storage projects is the subject of the first paper by Jhunjhunwala. The impact of intra-day storage projects is the subject of the second and third papers by Jhunjhunwala and Agarwal.

Hydropower is being promoted by the World Bank and other multilateral institutions because it supposedly helps reduce carbon emissions. The paper by International Rivers shows that the CO_2 and methane emissions from hydropower reservoirs are, in fact, contributing more to global warming than thermal power. Oxygen deep in the reservoirs is used up in the process of fermentation of inflowing organic matter. The water then becomes unaerobic. It has zero oxygen. All aquatic in this water dies. It becomes dead water.

The paper by Penny Schmitt tells us that unaerobic water is somewhat regenerated by infusing oxygen as it is released from the hydropower reservoirs. This provides further evidence that oxygen in the water is removed in the process of storage. This is the reason why oxygen has to be blown into the water.

These papers establish that, contrary to what the scientists commissioned by hydropower companies say, there is a huge negative impact of hydropower projects on the physio-chemical qualities of water.

7

Impact of Tehri Dam on Physio-Chemical Quality of Water

Bharat Jhunjhunwala

Editor's Introduction: Studies commissioned by hydropower projects claim that hydropower dams do not affect the chemical quality of water adversely. However, a close reading of data in the same studies indicates a negative impact. In this paper the author analyses data from four studies to establish negative impact of dams on physical and chemical qualities of water.

These studies relate to Tehri Dam where inter-seasonal storage of water is undertaken. Water is stored for many months in this dam.

Hydropower companies have commissioned studies by eminent academic institutions to study the impact of dam on quality of water. These studies often conclude that there in no negative impact of dams on water quality.

One crucial mistake common to the three studies is that impact is not measured between a without-and with dam scenarios. Comparison is made between the upstream and downstream quality of water. Such comparison tells us the impact of dam that is taking place *now*. It does not tell us what impact has taken place due to the construction of the dam.

Say, the Dissolved Oxygen was 8 mg/l at the site upstream before construction of the dam. It increased to 9 mg/l as the water flowed down the river to reach the site downstream of dam. The water gained oxygen @ 1 mg/l in that flow.

Now the dam was constructed. The DO level was same 8 mg/l at the upstream site. Now, it remained at 8 mg/l at the downstream site. Thus there was no change in DO. However, the river would have gained 1 mg/l oxygen if the dam had not been built. Therefore there was a loss of 1 mg/l due to construction of dam (Table 1).

Table 1: A Hypothetical Presentation of Impact of dam on Dissolved Oxygen (DO)

Sl. No.	*Situation*	*Upstream site*	*Downstream site*	*Impact of dam*
1.	After construction of dam	8	8	Nil
2.	Before construction of dam	8	9	–1

It is necessary to compare the water quality parameters before-and after making of the dam in order to arrive at a correct assessment of the impact of dam. The three studies do not make such an assessment. They only compare the *present* upstream-and downstream values. This only indicates the changes taking place presently. It does not indicate the change that has occurred due to making of the dam.

Even otherwise the data show impact of dams downstream when compared to upstream quality. In the following section I show from the original data that water quality downstream is poorer than upstream.

IIT Roorkee—AHEC

Ministry of Environment and Forests has got a study conducted by Alternate Hydro Energy Centre (AHEC) on the Cumulative Impacts of Hydropower projects on Ganga River. AHEC says:

> The water quality of... Tehri... and Vishnu Prayag have been compared with baseline water quality. The impact on DO is negligible; the BOD remains unchanged as the water passes through the tunnels/channels...The other parameters do not show significant change.

I have extracted the water quality data for u/s and d/s provided by AHEC for the Tehri project. Data for Vishnu Prayag is presented and discussed in the next chapter. The change noticed at Tehri is given at Table 2.

AHEC is correct in stating that impact on DO and BOD is negligible. However, change in the other parameters is severe as shown in last line of Table 2. Precisely these have not reported by AHEC. The statement that "other parameters do not show significant change" is totally unwarranted.

Secondly, AHEC mentions that water quality satisfies CPCB parameters for Class 'A'. This is correct. But CPCB parameters have been developed for drinking water purposes in an urban setting. They have not been developed for ecological assessment. For example, DO may decline from 12 mg/l to 6 mg/l. This will still satisfy CPCB Class 'A' requirement. Yet this hides a huge environmental impact. Many aquatic lives that require DO above, say, 10 mg/l may die.

IIT Roorkee—THDC

A study was done by Dr Pradeep Kumar of IIT Roorkee for THDC. No writeup on the study is available. However, data indicate a multifaceted impact of Tehri Dam on quality of water.

It is seen that 14 out of 19 parameters show a severe impact if 'severe' is considered to be 25 per cent or more (Table 3).

Table 2: Impact of Tehri Dam on Physio-Chemical Qualities of Water

Sl No	*HP*	*Temp*	*pH*	*DO*	*Cond*	*TDS*	*Turbidity*	*NO3-N*	*TP*	*BOD*	*Fecal C*
26	Tehri-u/s	11.7	7.8	9.43	169.9	80.4	0.71	0.8	0.98	1.87	20
	Chilyanisaur										
26	Tehri (Outlet)-d/s	14.1	7.9	9.26	113.4	53.6	5.77	0.4	2.66	1.92	240
	Change	-2.4	-0.1	0.17	56.5	26.8	-5.06	0.4	-1.68	-0.05	-220
	Change (%)	-20.51%	-1.28%	1.80%	33.25%	33.33%	-712.68%	50.00%	-171.43%	-2.67%	-1100.00%
	Severe?	Yes	No	No	Yes	Yes	Yes	Yes	Yes	No	Yes

Table 3: Water Quality as per IITR Study of Tehri Dam[1]

Sl. No.	*Parameter*	*Before Reservoir chilyanisaur*	*After crossing Tehri Dam*	*Change Impact Tehri Dam (Per cent)*	*Change Severe?*
1.	Fe (mg/l)	0.01	0.041	+ 310	Yes
2.	TDS (mg/l)	118	78	-33	Yes
3.	Conductivity (Mu/cm)	199	125	-37	Yes
4.	pH	7.95	7.63	-36	Yes
5.	Total Hardness $CaCO_3$ (mg/l)	90	60	-33	Yes
6.	Ca (mg/l)	24	15.2	-36	Yes
7.	Mg (mg/l)	7.2	5.3	-19	No
8.	Total Alkalinity $CaCO_3$ (mg/l)	64	45	-29	Yes
9.	Cl (mg/l)	1.2	1	-17	No
10.	SO_4 (mg/l)	35.9	21.1	-41	Yes
11.	Sodium (mg/l)	10	2.1	-79	Yes
12.	Potassium (mg/l)	3	2.3	-23	No
13.	PO_4 (mg/l)	0.031	0.036	+16	No
14.	Silica (mg/l)	8.3	6.2	-25	Yes
15.	DO (mg/l)	9.34	8.24	-12	No
16.	Turbidity	3.91	1.72	-56	Yes
17.	ORP (mV)	92.5	99.8	-8	No
18.	NO_3 (mg/l)	0.3	0.37	+23	No
19.	UV (m-1)	2.7	2	-26	Yes

National Environment Engineering Research Institute

NEERI has done a study titled *Water Quality of Bhagirathi / Ganga River in Himalayan Region.* The study has concluded "The levels of physico-chemical parameters including heavy metal (barring lead) concentrations did not indicate any adverse condition prevailing in the water samples".

Given below is table extracted from the same study.

Table 4: Physio-chemical Characteristics of Ganga Waters as per NEERI

Characteristic	*Average upstream*	*Average downstream*	*Change (per cent)*
Temperature (°C)	20.32	13.95	–31.37
pH	7.22	4.81	–33.45
Turbidity (NTU)	1.94	59.97	2991.24
Conductivity (us/cm)	111.10	76.45	–31.19
Suspended solids	25.60	54.55	113.09
Total dissolved solids	70.15	58.75	–16.26
($CaCO_3$)	42.90	45.11	5.15
Alkalinity			
Total hardness ($CaCO_3$)	38.23	36.33	–4.96
Ca-hardness ($CaCO_3$)	24.13	23.26	–3.63
Mg-hardness ($CaCO_3$)	14.59	14.33	–1.82
Dissolved Oxygen	9.56	5.79	–39.46
COD	4.13	2.69	–34.99
Total Kjeldahl Nitrogen	1.60	1.42	–11.56
NH_3-N	0.69	0.56	–18.98
NO_3	0.28	5.67	1925.00
SO_4	9.53	6.58	–31.01
PO_4-P	2.52	2.15	–14.88
Chlorides Cl	1.99	2.40	20.60
Na	2.27	2.60	14.32
K	3.14	1.26	–59.87

Source: Upstream figures are average of (6) Tehri upstream and (7) Bhilangana upstream. Downstream figures are average of (8) Tehri downstream 01 and (9) Tehri downstream 02.

Turbidity, suspended solids and nitrates show a huge increase. Dissolved oxygen shows a steep decline to less than 6, which is CPCB parameter for 'A' quality water.

The metal concentration as given by NEERI are given at Table 5.

Table 5: Metal Concentrations of Ganga Waters as per NEERI

Metal	*Average upstreamm*	*Average downstream*	*Change (per cent)*
Arsenic	0.019	0.063	228.9
Zinc	0.060	0.067	11.8
Lead	0.030	0.014	-55.0
Cadmium	0.007	0.008	23.1
Nickel	0.030	0.029	-5.0
Cobalt	0.019	0.063	240.5
Manganese	0.052	0.181	251.5
Iron	0.447	0.146	-67.4
Chromium	0.009	0.007	-17.6
Copper	0.011	0.003	-72.7
Mercury	0.008	0.001	-86.7

Source: Upstream figures are average of (6) Tehri upstream and (7) Bhilangana upstream. Downstream figures are average of (8) Tehri downstream 01 and (9) Tehri downstream 02.

It is seen that Chromium and Copper both show a decline. These are reported in the study to have beneficial impact on bactericidal properties of the water. Other metals, except zinc and nickel, also show a huge impact. I am not commenting on the direction of impact—whether increase or decrease—because either movement may have an ecological impact that is to be examined. The point is that its own data shows a large impact on metal content of water yet this is not reported.

Quality of sediments as given by NEERI is given at Table 6 and 7.

Table 6: Physico-chemical Characteristics of Ganga Sediments as per NEERI

Metal	*Average upstream*	*Average downstream*	*Change (per cent)*
Volatile matter	6.50	7.53	15.77
COD	13.56	5.16	-61.95
Metal	*Average upstream*	*Average downstream*	*Change (per cent)*
Sulphate	2.07	1.74	-15.98
Phosphate as P	3.74	1.32	-64.84
Calcium	0.97	2.03	109.28
Magnesium	3.93	1.75	-55.41

Source: Upstream figures are average of (6) Tehri upstream and (7) Bhilangana upstream. Downstream figures are average of (8) Tehri downstream 01 and (9) Tehri downstream 02.

Table 7: Metal Concentrations in Ganga Sediments as per NEERI

Metal	*Average upstream*	*Average downstream*	*Change (per cent)*
Arsenic	10.54	10.24	-2.85
Zinc	7.37	18.77	154.68
Lead	27.06	15.77	-41.72
Cadmium	4.45	6.06	36.18
Nickel	6.96	12.82	84.20
Cobalt	10.32	91.69	788.47
Manganese	135.71	1479.63	990.29
Iron	1812.95	1398.93	-22.84
Chromium	13.41	8.48	-36.76
Copper	23.95	6.70	-72.03
Mercury	1.23	0.40	-67.48

Source: Upstream figures are average of (6) Tehri upstream and (7) Bhilangana upstream. Downstream figures are average of (8) Tehri downstream 01 and (9) Tehri downstream 02.

Once again, like water, concentration of beneficent metals—copper and chromium—show a large decline.

PSI

People's Science Institute has undertaken a study on the impact of Tehri Dam on the quality of water. This study was

Table 8: Physio-chemical Characteristics of Ganga and its Tributaries as per PSI

Sl. No	*Station Code-Location*	*Temperature (°C)*		*Trans-parency*	*pH*	*Conductivity (μ mho)*	*Sat.*	*DOBOD (mg/l)*
		Air	*Water*					
4.	GBH04 u/s Tehri	17.0	16.0	NA	8.2	158	89.25	1.7
5.	GBH05 u/s Tehri	13.5	11.0	56	8.5	155	87.25	0.4
7.	GBH06 d/s Tehri	15.7	12.1	30	8.2	156	NA	NA

Table 9: Comparison of Key Parameters from Four Studies

Sl. No.	*Parameter*	*IITR-AHEC*	*IITR-THDC*	*NEERI*	*PSI*	*Result*
1.	Turbidity	0.7→5.7	3.9→1.7	1.9→ 60.0	Increase[2]	Increase (3/4 studies)
2.	pH	7.8→7.9	7.9→7.6	7.2→4.8	8.3→8.2	Decline—more acidic
3.	Conductivity	170→113	199→125	111→76	156→156	Decline
4.	Temperature water	11.7→14.1	NA	20→13	13.5→12.1	Decrease (2/3 studies)
5.	DO	9.4→9.2	9.3→8.2	9.6→5.8	—	Decline

undertaken in 2002-2004 towards the very beginning of impoundment in the dam. The results are given at Table 8.

The data indicate decline in transparency (or increase in turbidity), no change in pH and conductivity and DO, and a decline in BOD.

It is seen that turbidity increases, pH declines making the water more acidic, and conductivity and DO decline. There is a clear evidence of impact of dam on the chemical quality of water. Anil Gautam who was the study leader, has clarified that the results of pH, conductivity and DO may be at variance with other studies because this study was undertaken towards the very beginning of impoundment of the Tehri reservoir.

Comparison of Results

I give the main results from the four studies of Tehri Dam cited above at Table 9. Change in quality of water is seen in all their parameters.

Notes

1. IIT Roorkee has done studies at other time periods as well. However, I have been able to analyse the results of only one set of data for January 2007.
2. Figure given in this study is in terms of reduction of transparency from 56 to 30 cms which is tantamount to increase in turbidity.

8

Impact of Peaking Storage Dams on Water Quality

Bharat Jhunjhunwala

Editor's Introduction: Water is stored only for about 8 hours in intra-day or peaking storage dams. These are usually called 'Run-of-the-River' projects. However, this is a misnomer because flow of the water is held up even if only for few hours. Hydropower proponents say that such small period of storage does not impact the quality of water.

The author has extracted data for Vishnu Prayag hydropower project on the Alaknanda in Uttarakhand from a study undertaken by Dr Arun Kumar of Indian Institute of Technology, Roorkee to show that even such short duration of storage has a impact on quality of water.

The Ministry of Environment and Forests has got a study conducted by Dr Arun Kumar of Alternate Hydro Energy Centre (AHEC), Indian Institute of Technology, Roorkee on the Cumulative Impacts of Hydropower projects on Ganga River.[1] AHEC says:

> The water quality of… Vishnu Prayag have been compared with baseline water quality. The impact on DO is negligible; the BOD remains unchanged as the water passes through

the tunnels/channels... The other parameters do not show significant change (pp. 6-59).

I have extracted the water quality data for u/s and d/s provided by AHEC for Vishnu Prayag project. The change is given in the Table 1.

AHEC is correct in stating that impact on DO and BOD is negligible. However, change in the other parameters is severe. Precisely these have not reported by AHEC. The statement that "other parameters do not show significant change" is unwarranted.

Note

1. Study on Assessment of Cumulative Impact of Hydropower Projects in Alaknanda and Bhagirathi Basins upto Dev Prayag, 2011.

Table 1: Impact of Vishnu Prayag Project of Quality of Water as per AHEC

Sl. No.	*HP*	*Temp*	*pH*	*DO*	*Cond*	*TDS*	*Turbidity*	*NO3-N*	*TP*	*BOD*	*Fecal C*
1.	Vishnu Prayag u/s	6.6	7.6	8.92	81.4	38.3	1.2	2.14	0.93	1.8	39
2.	Vishnu Prayag d/s	8.9	7.8	9.04	192.8	94.9	1.1	3.4	5.4	1.6	21
3.	Change	-2.3	-0.2	-0.12	-111.4	-56.6	0.1	-1.26	-4.47	0.2	18
4.	Change (%)	-34.8%	-2.6%	-1.3%	-136.8%	-147.8%	8.3%	-58.9%	-480.6%	11.1%	46.1%
5.	Severe?	Yes	No	No	Yes	Yes	Yes	Yes	Yes	No	Yes

9

Impact of Hydroelectric Power Stations of NHPC on Water Quality

G D Agarwal

Editor's Introduction: National Hydroelectric Power Corporation (NHPC) is a major producer of hydroelectricity in India. NHPC has got a study done by National Productivity Council (NPC) of the Physico-chemical and bacteriological analysis at five of its hydroelectric stations. The study has concluded that there is no impact of the hydroelectric stations on water quality:

Baira Siul, HP: "It can be concluded that under the present conditions, the water quality at the project site has not been affected adversely due to construction of dam and reservoir."

Chamera, HP: "It can be concluded that under the present conditions, the water quality has not been affected due to construction of dam and reservoir."

Loktak, Manipur: "It can be concluded that under the present conditions, the water quality has not been affected due to construction of Ithai barrage or Power station."

Salal, J&K: "It can be concluded that under the present conditions, the water quality at all sites is almost the same and no adverse effect has been observed due to construction of dam and reservoir."

Tanakpur, UP: "It can be concluded that the water quality at the downstream of the reservoir and power station has not been affected adversely due to construction of dam and reservoir."

However, the data on the basis of which NPC has reached the above conclusions tells an altogether different story.

Like any other activity involving a river or the river-water, hydroelectric projects are also bound to make some impact on the water quality, though such impacts may be small in magnitude and hence may require more precise and careful measurements to be observed and substantiated. Most monitoring carried out at present is lacking such care and precision and generally assume from the very beginning that there is no impact. The result of such monitoring are so variable that any clear conclusions become difficult. This also applies to the sets of data for 5 NHPC Hydroelectric Power projects included in the 1994-95 Report of National Productivity Council, on which the present discussion is based.

Almost all hydroelectric projects involve some storage before diversion of the stream into a Head-Race channel or tunnel. Even the so-called "Run-of-the River" projects have at least a few hours (often 8 hrs.) storage caused by the obstruction in the path of natural flow. Obviously this change in velocity and detention is likely to result in some impacts on water quality. Subsequently the travel through the head-race channel/tunnel followed by the turbines and tail-race structures may also result in some impact. As such the present discussion shall study the impacts of the projects in two parts (i) Impact by obstruction and storage (ii) Impact of power-generation from inlet to head-race upto end of tail-race structure. The data are given at Table 1.

Impacts of Obstruction/Storage

Any detention is likely to cause some settling and result in some changes in turbidity and Suspended Solids (SS). Exposure to sunlight and consequent photo-synthetic activity shall result in changes in PH, alkalinity, DO which in turn could change

TDS and other chemical characteristics. Bio-chemical activity, particularly in case of long detentions could cause depletion of DO and changes in other chemical parameters. Such impacts are observed from the data at hand as per below:

Temperature: Rise of the temperature due to storage is best observed in case of Chamera project for the summer and after mansoon values and also for Salal project in all 3 seasons. Baira-Siul and Tanakpur projects also indicate slight increases in the temperature due to srorage, but Lok-tak project in Manipur state, where river temperatures were warm to start with, show small falls in temperature probably due to evaporative cooling.

Turbidity: The reductions caused by storage are obvious and large in case of Baira-Siul and Chamera, somewhat less in case of Salal and Tanakpur but get reversed to slight increases in turbidity in case of Loktak, most probably due to significant photosynthetic growth.

pH: Both trends are observed: (a) rise in pH due to photosynthetic activity-by 0.9 units in case of Loktak project summer value, by 0.4 units in case of Baira-Siul project after-monsoon and Chamera project winter value and 0.1 units in case of Chamera project after-monsoon value; (b)Fall in pH due to bacteriological decomposition of organic matter by 0.6 units and 0.8 units in the after monsoon and winter season for Loktak project, by 0.3 units in winter season at Salal and by 0.1 units in summer at Chamera, summer for Salal and after monsoon at Tanakpur. In the remaining cases, the two phenomena balance each other and no changes in pH is observed. The largest effects both for rise and fall in pH are observed at Loktak project where storage is large.

TDS: As in case of pH, both trends are observed: (a) Rise in TDS-at Baira-Siul (2 to 174 mg/l), at Chamera (3 to 20 mg/l) at Salal (4 to 7 mg/l) and Tankpur in winter (6 mg/l); and (b) Reduction in TDS at Loktak (34 to 39 mg/l) and Tanakpur after monsoon (5 mg/l).

Nitrates: Fall by 0 to 0.4 mg/l observed in most cases probably caused by bacteriological denitrification.

Table 1. Comparison of Water-Quality Characteristics at some NHPC Hydroelectric Projects as Studied by 1994-95 Report of NPC (The 3 values in a set denote the 3 seasonal values e.g. summer/after-monsoons/winters)

	Parameter	Units	*Baira-Siul Proiect*		*Chamera Project, HP*		*Loktak Project*		*Salal Project*		*Tanakpur Project*	
			Before	*After*	*Before*	*After*	*Before*	*After*	*Before*	*After*	*Before*	*After*
A.	**Comparison to Study Impact of storage /obstruction**											
(i)	Sampling Station	-	N^{o}l	N°3 Relevant	Mean of N°1& 2	N°4	Mean of N°4 &5	Mean of N°9 &10	N°1	N°5	N°1	N°3
(ii)	Temp	°C	10.4/12.2/10.2	10.4/14.9/	12.1/13.4/9.2	I5.3/16.O/9.0	28/23/17	27.7/22.6/16.5	14.9/13.9/0.9	15.6/14.7/7.9	21.0/15.8/11.6	-/I6.Q/I2.2
(iii)	Turbidity	NTU	60/19/40	4/3/-	26/8/21	2/6/2	3/4/2	5/2.5/7	106/56/32	94/45/23	200/51/56	-/40/49
(iv)	pH	-	7.7/7.6/8.0	7.7/8.0/-	7.9/8.3/8.0	7.8/8.4/8.4	8.1/7.9/8.0	9.0/7.3/7.2	8.6/8.0/8.4	8.5/8.0/8.1	8.1/8.5/8.3	-/8.4/8,3
(v)	TDS	mg/1	46/81/81	48/255/	71/123/121	91/132/124	97/87/110	72/53/71	80/110/54	84/114/156	142/173/160	-/168/166
(vi)	Nitrate	mg/1	0.9/0.5/0.8	0.9/0.1-	0.5/0.3/O.6	0.2/0.3/0.3	0.1/0.3/0.1	0.1/0.3/0.1	0.8/0.7/0.4	0.6/0.5/0.7	0.1/0.4/0.4	-/0.4/0.5
(vi)	DO	mg/1	9.2/9.1/9.6	9.5/7.9/-	9.9/9.4/10.9	8.2/9.2/11.5	5.8/7.5/8.5	5.7/7.7/7.7	10.6/10.5/11.6	9.9/10.2/11.3	9.0/9.9/10.4	-/9.8/10.4

DO: Both trends observed: (a) Rise due to photosynthesis at Loktak in after-monsoon (0.2 mg/l), Baira-Siul in summer (0.3 mg/l), Chamera in winter (0.6 mg/l); and (b) Reduction due to bacteriological decomposition at Baira-Siul in after monsoon (1.2 mg/l) Chamera in summer (1.7 mg/l), Loktak winter (0.8 mg/l) and Salal all seasons.

Impacts of Power Generation

These would include: (a) Deposition, dissolution as also biological/chemical activity in the head-race channel/tunnel depending on their length, material and velocity; (b) Escape of gases and corrosion/erosion of turbine and plant materials; and (c) Erosion of materials and dissolution of oxygen/CO_2 from atmosphere due to turbulence in the tail race structures. The following impacts are observed from the data at hand:

Temperature: Both trends are seen: (a) Rise in temp: 1°C in summer and 2°C in after monsoon at Baira-Siul, 0.5°C in winter at Loktak, 0.6°C in after monsoon and 1.6°C in winter at Salal; (b) Fall in temp: 1.3°C in winter at Baira-Siul, 6.2°C in summer, 1°C in after monsoon and 0.4°C in winter at Chamera, 0.7°C in summer and 0.6°C in after monsoon at Loktak, 0.3°C in summer at Salal, 0 to 2.7°C at Tanakpur. The rise would be due to conduction from warmer rocks around Head Race Tunnel while fall could be due to evaporative cooling at end of tail-race.

Turbidity: Small changes upto ±6 NTU in most cases, except large fall by 15 NTU in after monsoon values at Salal and rather large (14 to 36 NTU) increases in all seasons at Tanakpur. Small increases would be due to eroded/corroded material from structures/equipment and reductions due to dissolution of suspended solids in the turbulence of system. At Tanakpur there appears to be long scale erosion in Tail-race. The lone Salal after monsoon value appears to be an aberration in monitoring.

pH: Small changes upto ±0.3 units could be due to escape or dissolution of CO_2 in the turbine/tail-race systems. Large changes of –1.1 units in summer at Chamera and 0.7 and 0.5 units in summer and after monsoon at Loktak seem monitoring anomalies.

B. C omparision to Study Impact of HRT-PH-TRT

(i)	Sampling Station Relevant	-	$N^0 2$	$N^0 4$	$N^0 3$	$N^0 5$	Mean of $N^0 10\&9$	$N^0 7$	$N^0 5$	$N^0 7$	$N^0 2$	$N^0 4$
(ii)	Temp.	^0C	10.7/1l/8/8.1	11.7/13.86819	5/I7.3/9.2	13.3/16.3/8.8	27.7/22.6/16	15 27/22/17	15.6/14.7/7.9	15.3/15.3/9.5	23.7/16.1/11.8	21/15.5/11.8
(iii)	Turbidity	NTU	4/16/37	10/17/39	2/6/2	4/6/4	5/2.5/7	8/4/3	94/45/23	96/30/24	24/18/29	60/32'56
(iv)	pH	-	7.8/7.9/8.0	8.0/8.2/7.9	8.1/8.5/8.4	7.0/8.4/8.4	9/7.3/7.2	8.3/6.8/7.3	8.5/8.0/8.1	8.4/8.0/8.3	7.9/8.4/8.1	8.3/8.4/8.3
(v)	TDS	mg/1	48779/79	56/357/89	93/110/121	877100/119	72/53/71	71/56/62	84/114/156	88/128/153	163/178/159	155/170/163
(vi)	Nitrate	mg/1	15/0.5/0.8	0.8/0.4/07	0.1/0.3/0.2	0.1/0.3/0.1	0.1/0.3/0.1	0.1/0.1/0.2	0.6/0.5/0.7	0.9/0.6/0.8	0.1/0.3/0.4	0.1/0.5/0.5
(vii)	DO	mg/1	9.3/9.1/10.2	11.0/10/123	11.2/9.2/10.4	9.2/9.2/10.4	5.7/7.7/7.7	5.3/7.4/7.4	9.9/10.2/11.1	10.1/10.1/11.5	7.8/9.8,10.5	8.4/9.8.10.5

TDS: Small changes upto ±14 mg/l (mostly below 10 mg/l) on either side appear due to dissolution of structural materials or deposition/scaling on walls and surfaces. The large increase of 56 mg/l at Baira-Siul in after monsoon appears to be either large scale erosion in tail-race or anamoly in monitoring.

Nitrates: In most cases small reductions of upto 0.2 mg/p are seen but at Salal and Tanakpur increases of upto 0.3 mg/p are seen implying changes in oxidation power.

DO: Increases would be due to dissolution of atmospheric O_2 from turbulence in tail race, the largest being 1.7 mg/l in summer and 2.1mg/l in winter at Baira-Siul. The large fall in DO of 2 mg/l in summer at Chamera would be consumption in Head Race Tunnel.

Conclusion

The impacts of Hydroelectric Projects on water-quality may be small and on either side, but are surely not NIL. They may have implications on ecology and cannot be ignored.

10

Methane Emissions from Hydropower Dams[1]

International Rivers Network

Editor's Introduction: Organic matter such as twigs and leaves settle at the bottom of reservoirs. Here they ferment. The organic matter uses the oxygen available in the stagnant water to break CH compounds into CO_2 and H_2O. This leads to carbon emissions (of CO_2). Soon the oxygen stored in the water is exhausted but fermentation continues. After this CH compounds are broken into methane gas (CH_4). This gas adds many times more to global warming than CO_2.

The authors of this paper have extracted data from various studies of hydropower dams from across the world to show that the impact of dams on global warming is huge.

The emission of CO_2 and Methane gasses indicates that the water at the bottom of dams has less- or no oxygen and, therefore, becomes 'lifeless'. This is an obvious impact on water quality.

---

1. Adapted from: "Fizzy Science: Loosening the Hydro Industry's Grip on Reservoir Greenhouse Gas Emissions Research," International Rivers Network, 2006.

NEERI has estimated carbon and methane emissions from Tehri Dam at 2550 and 24 mg/km^2/day. These convert to about 52 grams CO_2/unit of electricity.[1] This is much less than the average of 2154 grams CO_2/unit estimated in this paper. The less CO_2 emission is mainly due to less surface area compared to dams referred to in this paper. Nevertheless, this establishes that methane is being emitted and quality of water released from Tehri is devoid of oxygen and is dead.

Introduction

The pages of a respected climate change journal are not a place one would expect to find a bad-tempered exchange over the merits of iconic soft drinks. Yet such a disagreement—over the rates at which Coca-Cola and Brazilian *guaraná* lose their fizz—was recently covered in the normally decorous pages of *Climatic Change*. While the immediate topic seems inconsequential to say the least, the larger context is of major importance/do tropical hydropower reservoirs cause greenhouse gas emissions to match those from fossil fuel plants?

Reservoir emissions research is highly politically charged and largely unknown by those outside the immediate field. Measurements have been made at only a tiny percentage of the world's half million or so freshwater reservoirs. Yet despite the

1. NEERI has estimated CO_2 and CH_4 emissions from Tehri Reservoir at 2550 and 24 mg/m^2/d respectively. NEERI has taken only the CH_4 emissions from surface. More CH_4 is released from degassing. IRN has given figures of Surface and Degassing for 3 tropical reservoirs from Brazil as follows: CH_4 surface = 0.035 Mt/yr; CH_4 degassing = 0.341 Mt/yr. Thus, degassing methane is about 10 times the surface methane. Taking this as a crude estimate we can assess the per unit emissions from Tehri.

   Surface CH_4 = 24 mg/m^2/d (as per NEERI). Thus, degassing CH_4 (crude estimate) = 24 x 10 = 240 mg/m^2/d (10 times the surface emissions). Total CH_4 = 264 mg/m^2/d. This can be converted to CO_2 equivalent = 264 x 25 mg CO_2 Equivalent = 6600 mg CO_2 taking methane as 25 times more powerful than CO_2 for global warming.

   Total CO_2 = 2550+6600 (2550 CO_2 directly emitted as estimated by NEERI + 6600 CO_2 equivalent of Methane as calculated above) = 9150 mg/m^2/d = 9.15 g/m^2/d = 9.15 x 365 g/m^2/yr = 3339.75 g/m^2/yr = 3339.75 x 42,000,000 g/yr (area of Tehri is 42 km^2) = 140,269.5 x 1000 x 1000 g/yr = 140,269.5/2700 g/unit = 51.9 g/unit (annual generation about 2700 million units).

preliminary nature of the science and its many complexities and uncertainties, available evidence strongly suggests that reservoirs are a significant global source of greenhouse gases, in particular methane.

There are a number of reasons for the low awareness of reservoir emissions even among policy makers and scientists working on climaterelated issues: the science is still relatively young, comparatively little has been published on it in peer-reviewed journals, and numerous uncertainties about net emissions levels remain to be resolved.

Danny Cullenward and David G. Victor of Stanford University wrote an editorial in *Climatic Change* in March 2006 in response to the Coke vs *guaraná* dispute. They note that most of the published work on reservoir emissions "comes directly from researchers connected to hydroelectricity companies, such as Eletrobrás or Hydro-Quebec." The Stanford analysts propose that a mechanism is needed to remove "any taint of interest" from reservoir emissions research. They believe that the best way to do this would be through a Special Report of the Intergovernmental Panel on Climate Change (IPCC). Cullenward and Victor note that the IPCC "is the only forum that could sustain the scientific integrity and transparency needed to synthesise the full international debate over emissions from hydroelectricity, as well as to make the information from such assessments available to climate policy makers in governments and international organisations."

The IPCC recognised in its 2006 guidelines on greenhouse gas inventories that reservoirs are a source of emissions, but also that more research is needed to be able to accurately quantify the extent of these emissions, especially of methane. It is a logical next step for the IPCC to take the lead in carrying out a comprehensive and independent review of research into reservoir emissions to inform their future guidance on measuring GHG fluxes.

Reservoir emissions were also recognised as an issue of concern in early 2006 by the governing board of the Clean Development Mechanism, the Kyoto Protocol's main carbon trading system. The CDMs Executive Board ruled that projects with large reservoirs relative to their generation capacity will not for the time being be allowed to sell carbon credits. The board's decision leaves open the possibility that they could in the future approve large reservoirs for carbon credits without adequately allowing for their emissions. This loophole should be closed and hydro projects with large reservoirs made ineligible for CDM credits.

The Executive Board distinguishes between hydro projects that may have high, medium or low emissions using largely arbitrary criteria based upon their "power density" (power generation capacity divided by area flooded). An IPCC Special Report could provide a scientific basis upon which the CDM could base its methodology for accounting for reservoir emissions.

The Carbonated Drink Skirmish and the Reservoir Emission Wars

The Coke versus *guaraná* skirmish was inadvertently started by ecologist Philip Fearnside of the National Institute for Research in the Amazon (INPA), in Manaus, Brazil. Fearnside, the world's second most-highly cited scientist on global warming in the decade up to 2006, is originally from the US but has been based in Brazil for the past 30 years. Writing in the September 2004 issue of *Climatic Change,* Fearnside used a comparison with the fizzing of a newly opened bottle of Coke to explain the massive surge of methane emissions that can occur by "degassing" when water is discharged under pressure at hydropower dams. According to Fearnside's calculations, degassing emissions from several large hydro dams in the Brazilian Amazon make these plants much larger contributors to global warming than fossil-fuel alternatives.

Fearnside has been in a long-running scientific feud on the scale of Brazilian hydropowers global warming impact with a team from the graduate engineering department (known as COPPE) at the Federal University of Rio de Janeiro. This team is led by Luis Pinguelli Rosa, who served from 2003 to 2005 as head of Brazil's Eletrobrás, one of the world's largest hydropower producers. Pinguelli Rosa and his colleagues also published an article in the September 2004 issue of *Climatic Change* which implied that Fearnside and others who argue that dams can have high greenhouse gas emissions are being seduced by the "lures of the thermo-power and nuclear power lobbies."

Pinguelli Rosa *et at.* published another *Climatic Change* article in 2005 responding to Fearnside's Coke analogy with the criticism that not all of the CO_2 in the bottle escapes immediately after it is opened. Bubbles, Pinguelli Rosa *et al.* point out, can be seen rising from a Coke bottle for "many minutes" after it is opened. Fearnside, they claim "seems unaware of empirical observation, clinging to his idealized convictions, whose theoretical grounds are certainly open to discussion." They also more generally (if some-what bizarrely) say that the use of the US drink is "highly symbolic of [Fearnside's] way of thinking."

The COPPE researchers put forward the popular Brazilian carbonated drink *guaraná* as a better tool to understand hydropower degassing. They note that *guaraná* is transparent while Coke is dark, so that its bubbles can be more easily seen, and people in Brazil "often sit around a table to chat as they drink it, with the bottles open and the glasses full for half an hour or more, without completely losing the bubbles. Instead of fast food, the Brazilian custom is a leisurely drink." More to the point, they say that the gas "does not escape in a few seconds from either the soda-pop bottle or the hydroelectric dam tailrace, as quite groundlessly affirmed by Fearnside." The COPPE team state that this supposed misconception of Fearnside's, together with "several scientific errors," explain why his calculation of emissions from Tucuruí, one of the largest reservoirs in the tropics, are several times higher than their calculation (see Table 1 and Figures 1 and 2).

Fearnside responded to the COPPE team with another article in *Climatic Change,* accusing Pinguelli Rosa and his co-workers of having "effectively made a career of trying to prove me wrong" since his first estimate of high emissions from Amazonian dams in 1995. Fearnside strongly defends the assumptions and conclusions of his previous paper and states that it makes little difference whether one assumes that dam degassing happens in 30 seconds or "the half hour they refer to for the last bubbles to emerge from a leisurely consumed bottle of Brazil's politically correct soft drink -*guaraná*. In either case, the high concentration of methane in the discharged water is emitted before there is time for bacteria in the river to degrade it to carbon dioxide, a much less potent greenhouse gas. Fearnside notes that "the longer this debate goes on and the more information that becomes available, the greater the [climate] impacts [of tropical hydropower dams] are found to be."

This spat over soft-drink culture is just the latest skirmish in the reservoir emission wars. These began in the mid-1990s when scientists working in the Canadian province of Manitoba published the first papers hypothesizing that reservoirs might be significant emitters of GHGs. Soon after, Pinguelli Rosa and Fearnside published their initial findings on Brazilian reservoir emissions.

The initial response from much of the hydropower industry to assertions that hydropower reservoirs were a source of GHGs on the scale of fossil fuel emissions per unit of generation was sputtering disbelief. A spokeswoman for the US National Hydropower Association responded to an IRN press release on reservoir emissions in 1995 with the statement that "It's baloney and it's much overblown... Methane is produced quite substantially in the rain forest and no one suggests cutting down the rain forest."

There are still those in the hydropower industry and elsewhere who continue to claim that reservoirs have no, or only negligible, climate impact, but most knowledgeable hydropower advocates now admit that, at least for some tropical reservoirs, GHG emissions may be substantial.

The assertions by Fearnside and the Manitoba scientists that hydropower could be a significant GHG source was of great concern to the hydropower industry, which was then beginning to market itself as "climate-friendly" and "zero-carbon." The huge utility Hydro-Quebec soon started to finance studies into reservoir emissions. In the second half of the 1990s, Hydro-Quebec began long-term collaborations on reservoir GHG studies with COPPE researchers financed by the Brazilian government and hydropower utilities.

Since 2000 the literature on hydropower emissions has grown relatively rapidly. It continues to be dominated by scientists in Brazil and Canada, with important contributions from scientists supported by Electricité de France who have published several papers on the Petit Saut reservoir in French Guyana. Two books of essays on reservoir emissions were published in 2005, one published by COPPE and the Brazilian utility Eletrobrás, the other a 730-page volume (from science publisher Springer) edited by Alain Tremblay and Louis Varfalvy of Hydro-Quebec and two Hydro-Quebec-sponsored academics. Neither volume contains contributions from Fearnside, nor from Eric Duchemin, a leading Québécois researcher on reservoir emissions who has remained independent of Hydro-Québéc. The International Hydropower Association, in which Hydro-Quebec plays an influential role, has published several fact sheets and articles based on the utility's reservoir emissions research and analysis. A booklet produced by IRN in 2002 is probably the only non-industry funded publication devoted to reservoir emissions.

The Hydro-Quebec researchers take a particularly fundamentalist line on the climate benefits of hydropower and often ignore not only Fearnside s research, but even that of their counterparts at COPPE. The manager of Hydro-Quebec's Greenhouse Gas Programme, Alain Tremblay for example, repeatedly claims that hydropower plants as a whole "generate 35-70 times less" GHGs per unit of power produced than fossil fuel plants. Such a conclusion ignores the science on tropical reservoirs. In addition to sponsoring researchers, Hydro-Quebec's efforts to control the terms of the debate have included cutting the funding of a team of Québécois reservoir emissions researchers, and trying to persuade an academic journal not to publish a peer-reviewed paper on reservoir emissions written by some of these researchers.

The main trend in reservoir emission science has been toward increasing recognition of the complexity of the pathways through which emissions are produced and released. Early work on the subject assumed that all emissions were fueled only by the biomass flooded when the reservoir was first filled, so that once this biomass had decomposed the emissions would

Table 1. Greenhouse Gas Emissions from Hydropower Plants

	Hydro plant	Power density (W/m^2)	Installed capacity (MW)	Flooded area (km^2)	CO_2 reservoir surface (Mt gas/yr)	CH_4 reservoir surface (Mt gas/yr)	CH_4 degassing (Mt gas/yr)	Total emissions (Mt CO_2eq/yr)	Electricity generation (GWh/yr)	Reservoir age (years)§	Emissions per kWh (gCO_2eq/kWh)
Boreal gross (Canada)	Sainte-Marguerite	10.38	882	85	0.02	0.000		0.02	2,770	N/A	8
	Churchill/Nelson	2.80	3,925	1,400	0.22	0.003		0.28	14,000	N/A	20
	Manic Complex	1.91	5,044	2,645	0.54	0.008		0.80	20,000	N/A	40
	La Grande Complex	1.20	15,552	13,000	3.28	0.039		4.10	82,000	N/A	50
	Churchill Falls	0.81	5,428	6,705	1.67	0.020		2.09	35,000	N/A	60
	Average	3.42	6,166	4,767	1.17	0.014		1.46	30,754	N/A	36
Tropical "reservoir net"* (Brazil)	Tucuruí	1.74	4,240	2,430	9.34#	0.094	0.970	31.56	18,030	6(1990)	1,751
	Curuá-Una	.56	40	72	0.04#	0.001	0.022	0.51	190	13(1990)	2,704
	Samuel	0.40	216	540	0.22#	0.010	0.030	1.06	530	12(2000)	2,008
	Average	0.90	1,499	1,014	3.20#	0.035	0.341	11.05	6,250		2,154
Tropical gross including degassing	Balbina	0.08	250	3,150	23.60	0.036	0.034	28.44	970	3 (1990)	29,322
	Petit Saut (French Guyana)	0.32	115	365	0.24	0.012	0.023	1.21	470	20 year ang	2,577
Tropical gross excluding degassing (Brazil)	Xingo	50.00	3,000	60	0.13	0.001		0.15	13,140	4-5	12
	Segredo	15.37	1,260	82	0.08	0.0003		0.09	5,519	6-7	16
	Itaipu	8.13	12,600	1,549	0.10	0.012		0.34	55,188	16-17	6
	Miranda	7.65	390	51	0.08	0.003		0.14	1,708	2-3	83
	Tucurui	1.74	4,240	2,430	7.52	0.097		9.55	18,571	14-15	514
	Serra da Mesa	0.71	1,275	1,784	2.59	0.033		3.28	5,585	3-4	588
	Barra Bonita	0.45	141	312	0.45	0.002		0.50	618	36-37	816
	Samuel	0.39	216	559	1.52	0.021		1.97	946	10-11	2,077
	Trës Marias	0.38	396	1,040	0.42	0.075		1.99	1,734	35-36	1,147
	Average	9.43	2,613	874	1.43	0.027		2.00	11,445	14-15	584

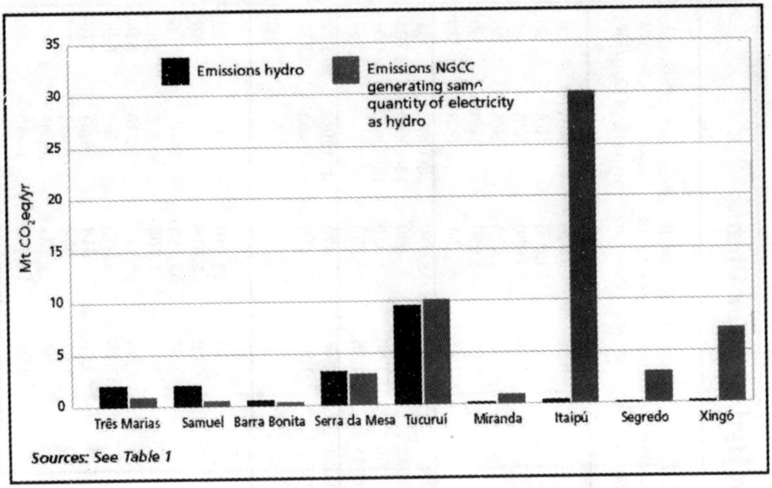

Fig. 1. Gross Tropical Hydro Emissions (Excluding Degassing)

stop. It is now understood that while there is a high initial pulse of emissions after reservoir filling, the emissions will continue to be fueled by carbon entering the reservoir throughout its lifetime. It was also believed that GHGs were emitted only at the reservoir surface, but it is now accepted that degassing at the dam can also be a source (although the actual scale of degassing emissions is hotly contested).

State of the Science: A Baseline of Agreement

Greenhouse gas emissions have now been measured at more than a hundred reservoirs, mostly in North America and Brazil. Despite the many methodological and analytical disagreements between independent researchers and those linked to the hydro industry, there are many points on which there is no serious dispute. These include:

All reservoirs can be presumed to produce methane (CH_4) and CO_2. Reservoirs are also sources of the potent greenhouse gas nitrous oxide (N_2O). A small number of reservoirs in boreal and temperate zones have been found to be sinks for CO_2 and N_2O (See Table 2).

The gases are released via diffusion across the water surface and in bubbles that rise from the reservoir bottom. There can also be significant emissions, especially at dams in the tropics, from the degassing of water released through turbines and spillways. When water from below the surface of the reservoir is discharged at the dam, the pressure acting upon it suddenly drops and—according to the chemical principle of Henry's Law-it is able to hold less dissolved gas. Degassing emissions are also due to the greater air/water interface created when water is pulverized at the bottom of a spillway or, as at Petit Saut, by a weir immediately downstream of the dam built to aerate the oxygen-depleted reservoir water and prevent it wiping out aquatic life downstream.

The major component of the warming impact of boreal reservoirs is diffusive CO_2; the major component of the warming impact from the *surfaces* of tropical reservoirs is methane bubbles. For at least some tropical reservoirs the majority of their warming impact is due to methane degassing (see Table 1).

The gases are formed by the decomposition in the reservoir of dissolved and particulate organic carbon. The main sources of this carbon—the "fuel" for the reservoir emissions—are the vegetation and soils flooded when the reservoir is first filled, the organic matter washed into the reservoir from upstream (which may be from natural or farmed ecosystems, or sewage from cities), the plankton and aquatic plants which grow and die in the reservoir, and the vegetation that grows on the "drawdown" land temporarily exposed during low reservoir periods. Reservoirs absorb atmospheric CO_2 due to photosynthesis by plankton and aquatic plants; this uptake can occasionally exceed CO_2 emissions. (See Fig. 3 for a graphic representation of the main factors influencing emission levels.)

Methane emissions occur due to bacteria that decompose organic matter in oxygen-poor water. The bottom layer of water in tropical reservoirs tends to be seriously depleted of oxygen. Some methane bubbles are oxidized to carbon dioxide as they rise to the reservoir surface-thus shallow tropical reservoirs where bubbles have less time to become oxidized tend to have the highest methane emissions.

Emissions per unit of area flooded are much higher from tropical reservoirs than from those in boreal zones, which are in turn generally higher than those in temperate zones.

Reservoirs emit greenhouse gases over their lifetime. There is an initial high pulse of emissions in the first few years after reservoir filling because of the huge amounts of carbon in the biomass and soils in the area flooded. Emissions generally appear to decline over subsequent decades. The actual rate of decline varies widely between individual reservoirs and climate zones. Some reservoirs fail to show any clear decline, and researchers have sometimes recorded increased emissions over time when sampling the same reservoir several years apart.

Emission levels vary widely between reservoirs depending upon such factors as the area and type of ecosystems flooded, reservoir depth and shape, the local climate, the duration of winter ice-cover, the area of the reservoir covered in aquatic plants, water quality (especially pH and nutrient content), the way in which the dam is operated, and the ecological, physical and socio-economic characteristics of the dammed river basin. Among the factors influencing degassing emissions are the concentrations of methane at different reservoir depths, the depth of turbine and spillway intakes, and the type of spillway design.

Surface emissions vary widely among different parts of the same reservoir (largely due to changes in depth, exposure to wind and sun, and growth of aquatic plants), and from year to year, season to season, and between night and day This greatly complicates efforts to develop reliable whole-reservoir estimates from a limited set of samples measured at specific points in the reservoir during specific time periods. Confidence in the measurements themselves is also hampered by the different results obtained through different measuring equipment and techniques, and disagreements over which measuring methods are most appropriate. Factors affecting degassing emission volumes include variations in the volume of water discharged, and the proportion of turbined water versus that which is spilled.

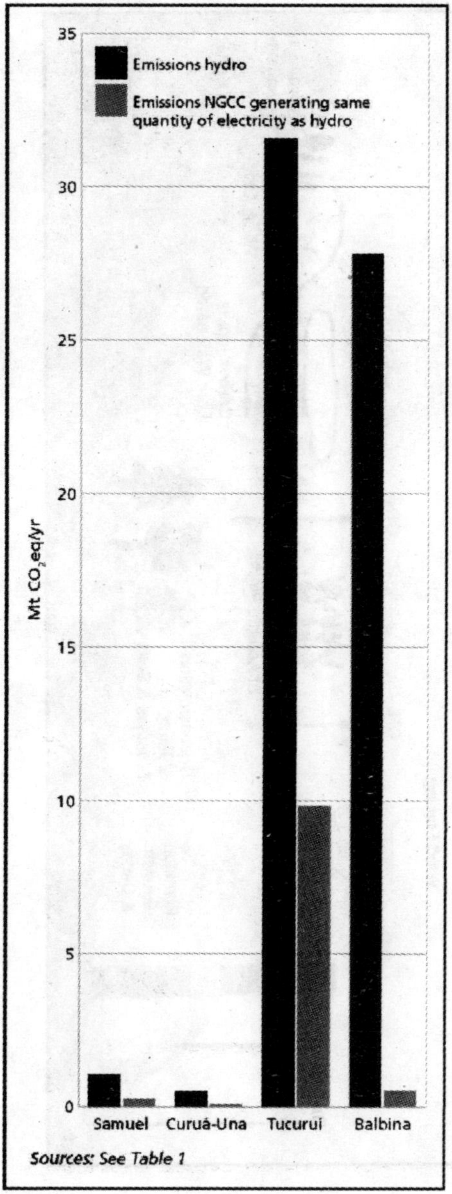

Fig. 2: "Reservoir net" Tropical hydro Emissions Compared to natural gas combined cycle generation.

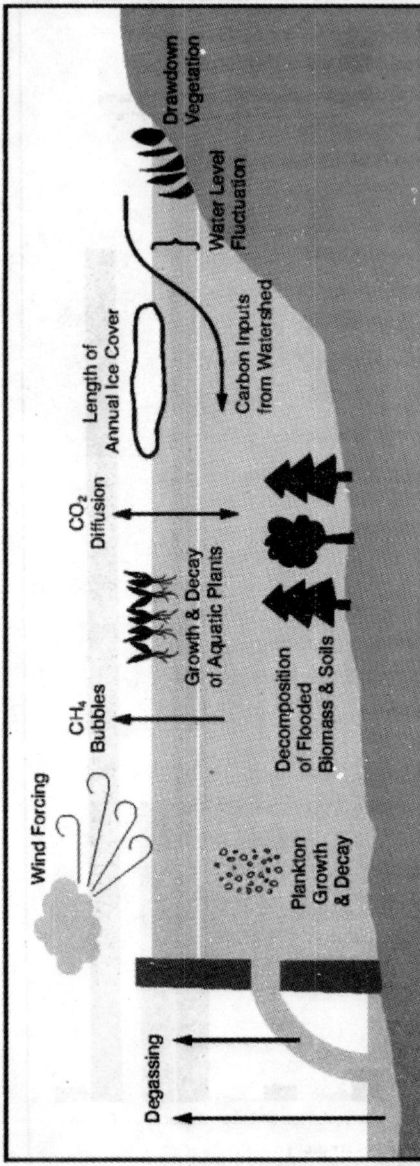

Fig: 3Some Key Factors Influencing Reservoir GHG Emissions.

Calculation of the warming impact of reservoirs should be based upon net emissions. This requires adjusting measurements of gross emissions at the reservoir surface and dam outlets to allow for whatever sinks and sources of greenhouse gases existed in the reservoir zone before submergence, the uptake of carbon through reservoir photosynthesis, and the impact of the reservoir upon the pre-dam flows of carbon throughout the wider watershed.

Degassing Emission

As the Coke vs. *guaraná* dispute shows, the other main area of disagreement among reservoir emission researchers is the magnitude of degassing emissions at tropical dams. Significant methane degassing emissions were first measured immediately downstream of the Petit Saut dam. These downstream emissions were much greater than the total volume of methane emitted from the surface of Petit Saut's reservoir. The only other tropical dam where degassing has actually been measured and the results published is Balbina, in Brazil.

Other Dam Impacts on Carbon Fluxes

A complete net life-cycle analysis of the warming impact of hydropower would need to consider various factors other than those considered in Fearnside's reservoir net methodology. However, at least for tropical reservoirs, including these factors would likely not require a significant adjustment to reservoir net emissions. This is largely because the impact of the reservoirs role as a factory for converting carbon into the powerful greenhouse gas methane (and in at least some cases for producing the even more powerful gas nitrous oxide) will likely far outweigh other impacts over the long term.

A life-cycle analysis of the climate impact of reservoirs should include the impacts of dam construction and decommissioning. Dam construction causes GHG emissions due to the use of fossil fuels by machinery and the production of building materials, in particular cement. Construction emissions vary according to the size and type of dams, with gravity dams (which contain a huge volume of concrete) having the highest emissions. Construction emissions could make up a significant

Table 2. Ghg Fluxes At the Air/Water Interface of Hydroelectric Reservoirs Worldwide, During the Open Water Period

	Climatic zone (size of sample)	*Mean age (year)*	*Range ($mg/m^2/day$)*	*Mean ($mg/m^2/day$)*
CO_2	Boreal (12)	24	653 - 2,500	1,459
	Temperate (16)	44	-1,195-2,200	525
	Tropical (21)	13	-142 - 13,737	5,467
CH_4	Boreal (7)	29	3.5 - 22.8	10.8
	Temperate (13)	49	1.3 - 15.0	6.7
	Tropical (22)	16	5.7 - 800	83.1
N_2O	Boreal (3)	18	0.02 - 0.5	0.2
	Temperate (0)			
	Tropical (4)	8	0.15-9.6	5.1
CO_2	Boreal (2)	10	1.0	1.0
	Temperate (1)	70	1.0	
	Tropical (16)	14	0.02-26	2.5
CH_4	Boreal (5)	20	0.04- 184.2	46.4
	Temperate (1)	70	14	
	Tropical (21)	15	0-800	85.6
N_2O	Boreal (2)	27	0 - 0.03	0.02
	Temperate (0)			
	Tropical (0)			

Blank entries = no data.
Source: Soumis et al. (2005).

Table 3. Estimated Greenhouse Gas Fluxes (mg/CO_2eq/m^2/day) for Different Climatic Zones and Ecosystems

		Boreal			*Temperate*			*Tropical*		
		CO_2	CH_4	N_2O	CO_2	CH_4	N_2O	CO_2	CH_4	N_2O
Forest	Mean	-863	-25	15	-1,455	-82	249	-2,317	-19	663
	Range	-5,230 – 271	-114 – 86	3-36	-16,605 –1,115	-263 – 8	-24 – 1,563	-5,926 – 0	-80 – 94	33 – 2,664
Marsh/	Mean		57			5,914	337	-5,688	4,513	83
Swamp	Range					-190 – 24,668		-11,634 – 3,443	-127 – 14,294	
Peatland	Mean	-744	2,131	1	703	8,488			11,132	
	Range	-2750 – 502	-38 – 12,270	-9 – 21	-540 – 2,400	607 – 22,074			8,096 – 16,445	
Lakes	Mean	1,304	2,621	20	1,886	489		-734	3,749	
	Range	-1,745 – 2,698	324 –3,518	-192 – 304	-1,116 – 2,921	31 –1,288		-1,232 – 52	1,219 – 7,544	

Blank entries = no data.

Sources:

Forest MarshiSwamp, Peatland - Tremblay et al. (2005b) Table 4.7.

Lakes –

Boreal (Canada) - Tremblay et al. (2005b) Table 8.1.

Temperate CO_2 (Southwest US) - Therrien et al. (2005) Table 9.1.

Temperate CH_4 (Wisconsin) - Bastviken (2004).

Tropical CH_4 (Venezuela) - Engle and Melack (2000).

Tropical CO_2 (French Guyana) - Tremblay et al. (2005b) Table 8.7.

component of the life-time emissions from a boreal dam, but would likely be insignificant compared to total emissions from a tropical project.

Dam decommissioning is likely to result in the mobilisation of a significant amount of accumulated sediments, potentially leading to a large pulse of carbon emissions. The growth of vegetation on the mudflats exposed by draining a reservoir, however, would be a carbon sink.

Emissions from human activities resulting from dam construction are potentially significant although these have not yet been calculated for any project (and would be very difficult to accurately ascribe to a particular dam). These secondary emissions include deforestation caused by farmers displaced to make way for the reservoir, and by fanners and developers clearing forests along access roads built for dam construction. Dams with an irrigation component may lead to increased methane emissions from newly watered farmland.

Emissions Overtime

Yet another layer of complexity when assessing reservoir emissions is how these may change over time. The usual assumption made about the pattern of reservoir emissions is that there is a large initial pulse of both CO_2 and CH_4 immediately after reservoir filling, followed by a gradual decline. At Petit Saut, methane emissions during the first year after filling the reservoir are estimated to represent around a quarter of the methane that would be released over the next century of the reservoir's life. It is not clear if N_2O emissions decline over time.

The actual rate of decline of GHG fluxes can vary widely between individual reservoirs. For many of the reservoirs studied, no obvious pattern of decline can be seen, especially for methane. Some of the highest methane emissions measured from a Brazilian reservoir came from a 36 year old project (Três Marias). The conditions for generating methane in tropical reservoirs continue over the life of the reservoir. Thus there is a strong theoretical basis for believing that after the initial pulse has subsided, substantial methane emissions will continue from a tropical reservoir until it is decommissioned.

Hydro-Quebec states that CO_2 emissions from boreal reservoirs fall to the levels of natural lakes found in the same basin within 10 years, and CH_4 levels within four years. A close look at Hydro-Quebec's actual data, however, reveals a different story: average emissions for 26 Québecois reservoirs ranging in age from 10 to 75 years show CO_2 emissions 50 per cent higher than average lake emissions; CH_4 emissions more than 14 times higher; and N_2O emissions almost twice as high. Other researchers in Quebec also report higher CO_2 emissions from reservoirs up to 70 years old than those from natural lakes. In any case, declining to natural lake emission levels does not mean declining to pre-dam levels: lakes can be substantial GHG producers and would normally be larger emitters than the ecosystems inundated by the dam (natural lakes cover around 2.3 per cent of the world's continental surface area yet are estimated to produce 6-16 per cent of natural methane emissions).

Recommendations

The governments that have ratified the UN Framework Convention on Climate Change (i.e. the Conference of Parties) should commission the IPCC to produce a Special Report on reservoir emissions. Cullenward and Victor point out that the 2005 IPCC Special Report on CO_2 capture and storage offers a useful model.

Although it is not obligatory to include methane emissions from reservoirs in GHG inventories to be submitted to the UN, countries with the scientific capacity should include these emissions using the methodologies in the relevant appendix of the IPCC's 2006 guidelines. Methane emissions should be included in the obligatory methodologies for the next version of the IPCC's guidelines (which hopefully would be based upon an IPCC Special Report).

Pending clarification of the link between power densities and specific emissions levels from the IPCC, the CDM Executive Board should amend its 2006 decision on hydropower emissions to rule that hydros with a power density less than ten are not eligible for credits.

11

Quality of Dammed Water Improves by Injection of Oxygen[1]

Penny Schmitt

Editor's Introduction: The author of this paper tells of improvement in quality of water released downstream of dams if fans of the turbines are fitted with air spray. The oxygen in the air is absorbed by water and quality of water improves. This improvement happens because the oxygen present in the water that impounded in dams is gradually used by fermenting organic material. The water has little or no oxygen. Thus injecting oxygen rejuvenates the water. This proves that water quality deteriorates in dams due to less oxygen.

Far below the deck of John H. Kerr Dam, dark green water boiled downstream. "There it is!" someone said. A host of white bubbles filled the water, turning it to a milky foam that spread out downstream of the dam toward Lake Gaston on the Roanoke River.

The white water gave visible proof that a recent retrofit (turbine venting) improves water quality and habitat for fish.

1. "Turbine venting improves water quality," http://www.hq.usace.army.mil/cepa/pubs/oct99/story7.htm

In turbine venting, a set of baffles pulls in and mixes air with water that is whirling through a turbine at up to 36,000 gallons per second. "Results at Tennessee Valley Authority (TVA) facilities show that turbine venting can add one to three milligrams per liter of dissolved oxygen," said Joe Tanner, Operations Manager at Kerr Dam. "Early testing results show that we're getting even greater benefits, in the range of 3.3 milligrams per liter. Our plan is to vent five more turbines before next summer."

"We're working to improve water quality on many fronts in North Carolina," said John Morris, Director of Water Resources for North Carolina. "We're delighted that the Corps has been proactive in finding a way to improve water quality in Lake Gaston, and making conditions better for fish and wildlife."

What's the big deal about such a tiny amount of dissolved oxygen? In this case, a little means a lot. In the summer, dissolved oxygen levels in water downstream of the Kerr Dam sometimes gets lower than is needed to support a healthy ecosystem. When oxygen gets low, all but the most tolerant fish species leave the upper reaches of Lake Gaston to find favorable conditions downstream.

"Reservoir tailwaters can be some of the most productive recreational fisheries resources available to the public," said Bud LaRoche, Regional Fisheries Manager for the Virginia Department of Game and Inland Fisheries. "The Kerr Dam tailwater is a very popular fishing area during the spring and early summer. However, during late summer and early fall, angling opportunities have been greatly reduced due to low dissolved oxygen levels. We applaud the Corps for taking these steps to improve water quality in the tailwater and upper reaches of Lake Gaston. The angling community should benefit greatly from this initiative."

Fish common to the tailwaters in good times include striped bass, largemouth bass, walleye, bream, crappie, and catfish.

Beginning last summer, the Corps worked with the Department of Water Resources and other interested North Carolina, Virginia and federal agencies, the Virginia Power

Company, and North Carolina State University to investigate other ways to improve oxygen levels. A 1995 TVA report on turbine venting convinced the Corps to consult with TVA to see if these procedures could work at John H. Kerr.

"We saw no reason why we couldn't repeat TVA's success," Tanner said. "This is a modification that could be accomplished within the Corps' current operating budget. We want to be able to say 'Yes, we can do something,' when our partners and neighboring communities ask for help. We're glad we found a way that we could take action."

To initiate the project, Tanner called on TVA's Jim Carter. "This is a procedure that only works on some turbines, "Carter said. "The turbines at Kerr are Francis turbines, and can be effectively retrofitted." Carter assisted the Kerr staff with pre-testing, design, installation of the first project, and post-testing. He trained Kerr staff so that they will be able to refit five more turbines next summer.

Total cost of the project will be about $150,000. "This year's effort cost $109,000," said Tanner. "We expect to be able to refit next year at a cost of between $5,000 and $7,000 per turbine, using our in-house work force."

"This isn't the first Corps of Engineers dam I've worked on," Carter said. "We've also been doing a similar project at Hartwell Lake in Savannah District. We've seen a lot of success at our own facilities and at others, when conditions are right. I encourage any facility that is looking to improve dissolved oxygen levels to determine whether this method will work for them."

"Turbine venting is the most economical and simple measure that can be taken to add oxygen to the river," said Chuck Wilson, a biologist in Wilmington District. "It's the logical first thing to try."

Wilson explained the conditions turbine venting will help alleviate. "During the hot months, deep lakes develop thermal stratification," he said. Simply put, that means there are stable heat differences between the top layer of the lake exposed to hot summer sun, and the coolest bottom layer of the lake.

When this layering occurs, warm water on the surface does not mix with the cooler bottom water. "Decomposing plant and animal material and microscopic plankton living in the bottom waters cause this layer to become 'anoxic,' that is, without oxygen," Wilson said. "Thermal stratification is a natural process in deep water lakes, and is not a problem with the Kerr Reservoir. However, because the intakes are in the bottom layer, when water is drawn through the dam to operate the turbines and generators, the bottom layer goes first. Turbine venting pulls air into this water, adding needed oxygen for the fish that live downstream in Lake Gaston."

"Now that we know the first turbine vented at Kerr Dam shows favorable results, our staff will modify all the remaining large-size turbines at the dam," Tanner said.

"This is a winning project for everyone," Morris said. "With the Corps' help, we can improve water quality on the Roanoke River and Lake Gaston. Our fisheries will improve, people visiting North Carolina and our sister state, Virginia, will find better outdoor recreation opportunities, and all this has been done at the lowest possible cost to taxpayers."

PART 4: IMPACT OF DAMS ON BIOTIC QUALITY OF WATER

Introduction

One indicator of health of the river water is the aquatic life that it supports. A healthy river water supports greater biodiversity than unhealthy water.

The three papers in this section provide evidence of the negative impact of hydropower projects on aquatic biodiversity.

12

NEERI Study on Coliphage[1]

Bharat Jhunjhunwala

Editor's Introduction: Tehri Hydro Development Corporation (THDC) has asked National Environment Engineering Research Institute (NEERI) to undertake a study of the Self-Purification Capacity of River Bhagirathi: NEERI has concluded that the Tehri Dam will not have any adverse impact on the self-purifying capacity. This capacity arises from the coliphages that are adsorbed in the sediments. The supply of sediment from Tehri Dam and from Alaknanda River that joins downstream at Dev Prayag will be adequate to maintain the self-purifying capacity. However, NEERI fails to examine the fact that these sediments will neither be created nor adsorbed after most of the river is diverted into tunnels or reservoirs. A single project like Tehri may not have much negative impact but a cascade of dams will most likely destroy the self-purifying capacity.

Executive summary of the 2004 study is reproduced in the first part of this article. It is followed by comments from the author.

1. This article is based upon comments made to NEERI by the author and Shri G D Agarwal. Shri Agarwal's assistance is thankfully acknowledged.

Executive Summary of NEERI Study, 2004

The Tehri Dam project envisaged to capitalize on the 2328 sq km snow bound area and on the rainfall ranging from 1016-2630 mm, in the total catchment area of 7511 sq km of river Bhagirathi up to Tehri. The Tehri Dam has the capacity to store 2615 million cumecs, out of the total discharge of around 7400 million cumecs of the Ganga valley. The dam is also likely to capture around 65 per cent of the total sediment carried by Ganga at present. Tehri Hydro Development Corporation Ltd., Rishikesh retained National Environmental Engineering Research Institute, Nagpur in 2002 to investigate the impact, if any, of the dam being constructed at Tehri, on the self purification capacity of the river Bhagirathi during post-commissioning phase.

Twenty sampling points were identified in the stretches from Gomukh to Rishikesh and from Badrinath to Devprayag. Maximum five sets of samples were collected from identified sampling points of the rivers (Bhagirathi, Bhilangana, Alaknanda, Mandakini and Ganga) during September 2002 to August 2003 for assessment of different abiotic, biotic and microbiological parameters.

Application of water quality index (WQI), which is based on nine parameters, viz. DO, pH, BOD, temperature, total solids, turbidity, total-P, NO_3 and fecal coliform, at various stretches of the river water revealed that water quality of the river between Tapovan and downstream of Uttarkashi was good (index value 70–90) throughout the period of study, while water samples of other areas were ranging medium and good during the same period. An exercise further revealed that WQI values would have been excellent (90–100) throughout the study area, if fecal coliform values were eliminated from the nine parameters used for WQI calculations.

Based on the irrigation water quality classification, water quality of all the sites of the entire stretch was determined and found to fall under the desired category of C_1S_1. The irrigation quality of water having C_1S_1 category is beneficial for growing plants like: *Eucalyptus robusta*, *Acacia nilotica*, *Casuarina* sp., *Prosopis tuliflora*, *Dal-bergia sissoo*, *Azadirachta indica* etc.

It has been observed that water samples from different rivers contained specific types of phages. Different types of hosts were required for their detection. The experimentation done at NEERI revealed that Ganga/Bhagirathi sediment has the capability to adsorb coliphages and induce their proliferation. The coliphages adsorbed to the sediment appear to be responsible for pre-dating coliforms in the overlying water column when the sediment and water co-exist in a container under static condition. The Bhagirathi water to be stored in a dam mimics such static condition and, therefore, shall in no way deteriorate the water quality of the river downstream of the Tehri dam.

The release of copper and chromium under static condition and the synergistic effect of chromium on the bacteriostatic/ bacteriocidal property of copper appear to be the factors which keep the water free from coliforms and other bacteria responsible for putrification of water when left for a long time under static condition.

Quantification of U_3O_8, ThO_2 and per cent K in sediment samples and comparison of these parameters with those present in other river sediment samples and freshwater lake sediment samples show that Bhagirathi/Ganga sediments collected between Gomukh and Rishikesh are more radioactive than others though it could not be established as to whether the radioactivity observed could be bacteriocidal. However, possibility of existence of synergestic effect of radioactivity on the antibacteriocidal activity of Cu and Cr, in combination, cannot be ruled out.

It can be concluded now that the uniqueness of river Bhagirathi/Ganga lies in its sediment content which is more radioactive compared to other river and lakewater sediments, can release Cu and Cr which have bactericidal properties and can harbour and cause proliferation (under static condition) of coliphages that reduce and ultimately eliminate coliforms from the overlying water column. This is possible as the dam is going to retain practically all the sediment load of Bhagirathi as particles of size of >0.01 mm are likely to be retained in the dam. Thus, Tehri dam is not likely to affect the quality or self preservation property of river Bhagirathi/ Ganga, as it mimics a static container which is conducive for conditions responsible to maintain the water quality.

Comment

Preservation of Coliphages

The NEERI study has homed in on coliphage as the source of the self-purifying capacity of River Ganga. There are two types of bacteria in the river water—coliform are harmful bacteria while coliphage destroy the coliform and are beneficial bacteria. Usually a particular species of coliphage destroys one particular species of coliform. There are innumerable species of both coliphage and coliform. Speciality of Ganga lies in the fact that the coliphages are 'wide-spectrum'—one coliphage destroys many species of coliform.[1] This helps the river self-purify itself more easily.

The coliphages are adsorbed in the sediments after having been created. They lie dormant in the sediments—even for many years—and get revived when coliform enter the water. NEERI has found that coliphages are present in the river water downstream of Tehri Dam up to Rishikesh. Moreover, NEERI has assessed that 10 per cent release of sediments from Tehri Dam, plus sediments being added by Alaknanda River at Dev Prayag, will be sufficient to supply the sediments required for the coliphage to survive and multiply hence there is no negative impact of Tehri Dam on the self-purifying capacity of the Ganga.

The special quality of Ganga waters is also due to minute (but high) levels of Thorium, which is radioactive, and high levels of copper and chromium in its water.

These beneficial elements—thorium, copper and chromium—enter the river water from the rocks against which the water rubs during flow:

> The metal ions, originally derived from breaking of rocks, are controlled by several river valley conditions—physical as well as chemical (NEERI 2004: 74; NEERI 2011: iii).

This adsorption takes place in two ways—chemical- and mechanical weathering. Rain water contains miniscule amounts of acid. This acid breaks the rocks and loosens the metals

1. This was mentioned by scientists of NEERI during personal discussion at NEERI on 18 February 2010. This is not mentioned explicitly in the NEERI report.

therein which then flow into the river. This is called chemical weathering. Secondly, metals are directly absorbed in the river water as the water dashes against the rocks containing these metals. The difference in metal composition of the river is said to be due to this difference in the weathering regime:

> The striking feature of the radium isotope data is the distinct difference in the ^{228}Ra and ^{226}Ra abundances between the highland and lowland rivers. The lowland waters are enriched in ^{228}Ra, while the highland waters contain more ^{226}Ra. This difference mainly results from the differences in their weathering regimes (NEERI 2011: 169).

The problem is that if upstream river is diverted into tunnels for generation of hydropower, then the water will not rub against the stones and not absorb the beneficent metals. It is to be noted that bumper-to-bumper dams have already been made upstream of Tehri Dam except the uppermost 135 km; and are planned on the entire 300 km flow of Alaknanda except the lowest 30 km stretch. Thus, it is likely that the coliphages will be deprived of the sediments on which they survive. The metals loosened by chemical weathering due to the rains may seep into the dry river bed and also not get carried by the river.

Conclusion is that the single project of Tehri Dam may not adversely affect the sediments because the sediments have already been created during the upstream flow or are being added from other rivers. However, a cascade of projects which prevents weathering in upstream reaches and other rivers will prevent adsorption of the beneficent elements in water and remove the base on which the coliphages survive. The conclusion that Tehri Dam will not affect the self-purifying capacity of Ganga waters may possibly be correct on a stand-alone basis; but this cannot be extrapolated to other projects especially when a cascade is being built which will almost totally divert the upstream river waters into tunnels and prevent weathering.

Creation of Coliphages

The NEERI study does not give any indication of the reasons of creation of the wide-spectrum coliphages. There is an oblique suggestion that this may be due to the miniscule amounts of beneficent radioactivity on the river water. However, this is not substantiated. If this were so, it should be possible to replicate their creation under laboratory conditions by exposing water of other rivers to same small levels of radioactivity. NEERI agreed to this suggestion for further research but expressed inability to undertake this in absence of another sponsored project.

It is possible that the vibrations of sages in the area may be leading to the creation of these coliphages. Or a unique combination of flow velocity, weathering, algae, temperature etc. may be leading to their creation. It is wise to apply the 'precautionary principle' until we know the precise conditions under which these special coliphages are created. It is best to leave the river in its pristine conditions till then.

Algae

River water has small micro-organisms that are food for bacteria. These are called 'phytoplankton'. The quality of aquatic life substantially depends upon the availability of variety of phytoplankton. NEERI has calculated the diversity indices of phytoplankton in Ganga waters. These are reproduced below:

Table 1: Average Density, Diversity and Composition of Phytoplankton at different Sampling Points of the Rivers/ Tributaries in the Study Area (September, 2002 to August, 2003)

Sl. No.	*Sampling Location*	*Counts/ml*	*Shannon Weiner Diversity Index (SWI)*	*Palmer's Index*
7.	Tehri upstream (Bhagirathi)	0	0	0
8.	Tehri upstream (Bhilangana)	46	1.55	2.0
9.	Tehri downstream (Bhagirathi)	22.75	0.68	0.75

Source: Table 17, NEERI 2004.

A reading of above table shows that the Tehri upstream shows complete absence of phytoplankton and diversity Index of 0. This may be due to upstream hydropower projects which cause almost all the water of the river to flow through tunnels, deprive microorganisms of sunlight and air that are necessary for their development.

The table also shows a major decline in the count as well as the diversity index between upstream Tehri (Bhilangana) and downstream Tehri (Bhagirathi). These data indicate a negative impact of Tehri Dam on microorganism. NEERI has not assessed the implications of this decline.

References

NEERI (2004), *Self-purification Capacity of River Bhagirathi: Impact of Tehri Dam*, NEERI, October 2004.

NEERI (2010), *Reply to queries from Dr. G.D. Agarwal and Dr. B. Jhunjhunwala (April 05, 2010 by e-mail).*

NEERI (2011), *Water Quality of Bhagirathi / Ganga River in Himalayan Region*, NEERI, July 2011.

13

Impact of Tehri Dam on Aquatic Biodiversity[1]

R C Sharma, Punam Chauhan and Manju Bahuguna

Editor's Introduction: One indicator of health of water is the biodiversity of aquatic life present in it. Good quality of water will be able to sustain larger number of species of organisms because it can meet their requirements. Poor quality of water will be able to sustain fewer species of organisms. A reduction in aquatic biodiversity, therefore, is indicative of deterioration in quality of water.

This study shows that biodiversity has been reduced downstream of the Tehri Dam. This indicates deterioration of water quality.

Introduction

Humans started to build dams and canals a long time ago and in this way influenced the natural flow regime. In 2600 BC, the first large dam (14 m height, 113 m crest length) was built

1. Adapted from "Impact of Tehri Dam on Aquatic Macroinvertibrate Diversity of Bhagirathi", Uttarakhand (India), *Journal of Environmental Science and Engineering,* Vol. 50, No. 1, Page 41 50, January 2008.

at Sadd el Kafara in Egypt in Garawi ravine facing Memphins[1]. Dam construction and the dam itself have variable impacts–social, economical, geographical-on water quality, climate, flora and fauna[2]. Tehri Dam, which is one of the five biggest dams of the world and biggest in Asia, has attracted the attention of environmentalists, geologists, seismologists, social scientists and fishery biologists. In order to protect the aquatic ecosystem, it is important to understand the health of the stream or river. One method of assessing river health is by analysing benthic organisms, which are visible animals that live on the bottom of aquatic environment. Such organisms, which are primarily insect larvae and aquatic worms, can include an entire array of life. Benthic communities are ideal for studying aquatic ecosystems, because they are mostly sedentary and have long life cycles. These biological forms of aquatic system act as bioindicators whose presence, absence or condition provides information about environmental requirements for them to be healthy and reproduce successfully. The presence or absence of healthy population of organisms within their habitats is a sign of particular environmental characteristics. In streams and rivers, potamologists often look for macroinvertebrates to get an idea of water quality. Macroinvertebrate community has been considered as an indicator of stream ecology[3]. The macroinvertebrate population is very sensitive to any environmental perturbation and is highly influenced by environmental change. Distribution, density and biomass of benthic organisms depend upon the physico-chemical characteristics of the water, the nature of the sediments or substratum, biological complexes such as food, predation and other factors.

There are many studies describing the changed flow and temperature regimes, channel morphology, water quality, quantity and quality of paniculate matter and their influence on tailwater bottom fauna[4-5]. But only a few studies have described detrimental effects in quantitative terms[6] together with determination.

A bulk of contribution on the macroinvertebrates has been made[7-9]. Some of the studies have also been made on macrozoobenthos in different parts of India[10-13]. However, no study on the impact of Tehri Dam construction on aquatic macroinvertebrates has been carried out except few scientists[14, 15]. Thus, it was found necessary to provide in-depth information on the impact of construction of Tehri Dam, one of the most controversial dams on macroinvertebrate diversity of Bhagirathi.

Study area

Tehri Dam is constructed across the river Bhagirathi, downstream the confluence of Bhagirathi and Bhilangana at Tehri (latitude 30°23' N; longitude 78°29' E), Uttaranchal, India (Fig. 1). Bhagirathi is one of the parent streams of Ganges and originates from the Gangotri glacier of Garhwal Himalaya. Tehri Dam is 260.5 m high rock-fill structure designed to produce 2400 MW of electricity and to irrigate 2.70×20^5 hectare land.

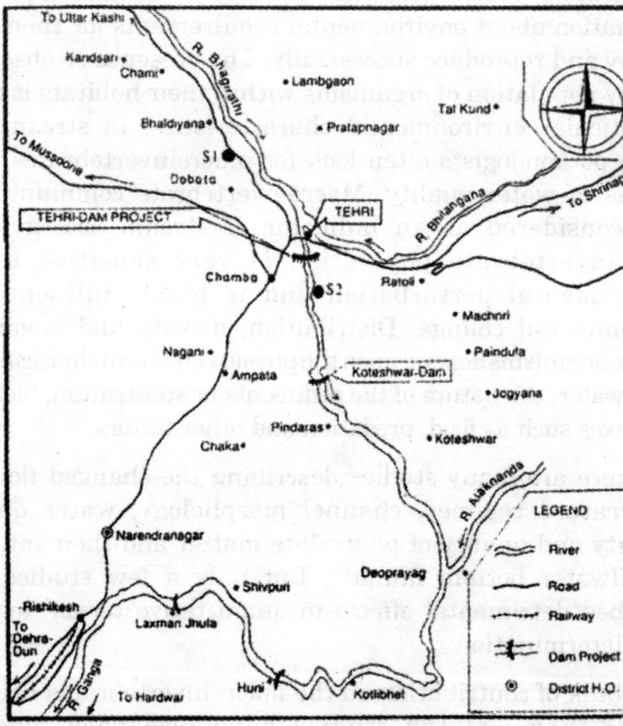

Fig. 1. Location map of study area.

Although, its construction will provide lots of benefits to the country but the process of construction is responsible for the destruction of a vast area of terrestrial and aquatic ecosystem. Various construction activities (construction of approach roads, adits to powerhouse, digging of tunnels, construction of coffer dam, etc.) have drastically transformed the aquatic ecosystem of Bhagirathi. Two sampling sites, i.e. the reference site (S_1) which has no impact and the site (S_2) which has the impact of construction activities, were identified. Ecological characterisation and composition of bottom substrates of both the sampling sites (S_1 and S_2) are presented in Table 1. The bottom substrates at the sampling site S_1 were represented by large and small cobbles, and the embankment was stable with rich riparian vegetation. However, the bottom substrates at the impacted site (S_2) were dominated by big boulders with a few

Table 1: Physical habitat indicators, river chemistry indicators and biological indicators of the sampling sites (S_1 and S_2) of the river Bhagirathi, the Tehri Dam area, Uttaranchal, India

Indicators of stream conditions	S_1 *(Sirain)*	S_2 *(Kyari)*
Physical habitat indicators		
Elevation (m above m.s.l.)	640.0	635.0
Mean air temperature (°C)	21.75	22.42
Embankment vegetation	Good riparian	Degraded
Boulders (>256 mm) (%)	30	73
Cobbles (64-256 mm) (%)	36	20
Pebbles (16-64 mm) (%)	24	2
Gravels (2-16 mm) (%)	9	
Sand (<2 mm) (%)	1	5
River chemistry indicators		
Mean water temperature (°C)	14.75	15.61
Mean velocity (m sec^{-1})	1.87	2.57
Mean dissolved oxygen (mgl^{-1})	10.69	9.71
Biological indicators		
Density (ind. m^2)	943.8	166.7
Diversity indices (H')	3.44	1.98

cobbles and pebbles. Embankment of the impacted site was degraded and eroded by several activities of Tehri Dam construction.

Materials and Methods

A detailed survey of the study area was undertaken to identify the sampling sites. Two sampling sites (S_1 and S_2) were identified to determine the diversity, composition and abundance of aquatic macroinvertebrates in addition to the physico-chemical characteristics of water. The sampling site S_1 was selected as a reference site at Sirain, 5 km upstream the Tehri Dam, however, the another sampling site S_2 was selected as the impacted site at Kyari, 5 km downstream the Tehri Dam on Bhagirathi river. Monthly samplings for analysing the hydrological parameters and biotic components of macroinvertebrates were conducted for one annual cycle (September 2004 to August 2005). Temperature of air and water was recorded with the help of a centigrade (0-110°C) thermometer. The mean velocity of water was measured using the Electromagnetic Current Meter (Model-PVM-2A). The vertical velocity profile was established by measuring the velocity at depths ranging from 0.10 m to 1.5 m above the substratum. pH of the water was determined by the Control Dynamics pH meter (Model-APX15/C) while turbidity was measured by the Turbidity Meter (Model-5D 1M). Phosphates were determined with the help of the Systronics Spectrophotometer. Chlorides, alkalinity and hardness were calculated by titration methods following the standard methods outlined in APHA[16].

The percentage cover of different-sized substrata within each surber quadrate was estimated visually using the substrate size classes of sand (0.06-2.0 mm), fine gravel (2-32 mm), coarse gravel (32-64 mm), cobbles (64-256 mm) and boulders (>256 mm)[17].

Benthic invertebrates colonizing the bottom substrates were sampled with the surber sampler (0.5 mm mesh net) to a depth about 10 cm in a quadrate. Samples were preserved in 4 per cent formalin. These invertebrate fauna were identified to the possible lowest taxonomic level and counted at the laboratory. The qualitative and quantitative analyses of water current macroinvertebrates were made by using the standard methods[16,18]. Identification of these aquatic organisms was made with the help of several taxonomic keys of Freshwater Biological Association, UK. The density of macroinvertebrates was calculated for all the five replicates collected from each site and then their mean was calculated.

Statistical treatment of data

Species diversity indices were calculated using the Shannon and Wiener diversity index[19]:

$$\text{Diversity index } (\overline{H}) = \sum_{1=1}^{s} \left(\frac{n_i}{N}\right) \log_2 \left(\frac{n_i}{N}\right)$$

where,

$\overline{H}$ = Shannon Wiener index of diversity;
n_i = total no. of individual of a species;
N = total no. of individuals of all species.

Karl Pearson's correlation coefficient was used to know the relation of hydrological attributes among themselves. Multiple regression analysis was also performed between the density of macroinvertebrates and hydrological attributes were determined, to know the influence of hydrological attributes on biological communities.

Results

Hydrological attributes

The physical habitat indicators, chemistry indicators and biological indicators of Bhagirathi river at the reference site (S_1) and impacted site (S_2) have been presented in Table 1. The reference site S_1 is covered with good riparian vegetation and comprises of 30 per cent boulders, 36 per cent cobbles, 24 per cent pebbles, 8 per cent gravels and 2 per cent sand, while the impacted site S_2 has degraded embankment and comprise of 73 per cent boulders, 20 per cent cobbles, 2 per cent pebbles and 5 per cent sand. This is the clear indication that the natural composition of bottom substrates at S_2 has been drastically altered. Bottom substrates provide natural refuge

Table 2: Monthly Variations in Physico-chemical Parameters of Bhagirathi River Recorded at the Reference site (S1) for the Period September 2004 - August 2005

Parameters	*Sep*	*Oct*	*Nov*	*Dec*	*Jan*	*Feb*	*Mar*	*Apr*	*May*	*June*	*July*	*Aug*	*X±S.D.*
Air temperature (°C)	27.5	21.0	15.0	16.0	14.0	18.0	19.0	20.5	25.0	26.5	29.0	29.5	21.75 ±5.57
Water temperature (°C)	16.0	14.0	10.0	10.0	9.5	11.0	15.5	20.0	19.0	18.2	17.0	16.8	14.75 +3.77
Water current (m sec^1)	2.5	2.35	2.08	1.96	0.93	0.95	0.62	0.99	1.86	2.08	3.07	3.05	1.87+.0.83
HMD(m)	1.75	15	0.4	0.54	0.46	0.4	0.36	0.94	1.0	1.26	2.0	2.8	1.12 ±0.77
Turbidity (NTU)	5	8	4	3	6	10	18	92	10	218	501	798	139.42+254.28
Transparency (m)	0.14	0.8	0.4	0.54	0.46	0.4	0.15	0.32	0.15	0.06	0.04	0.02	0.29 ±0.24
Conductivity (µS cm^1)	0.12	0.15	0.19	0.21	0.18	0.16	0.13	0.103	0.109	0.093	0.01	0.044	0.13 ±_ 0.06
TDS (mgl')	12	8	42	12	2	2	22	42	26	46	88	84	32.17 ±29.46
Dissolved Oxygen (mg 1^1)	10.03	10.05	10.94	11.35	11.76	12.16	11.55	10.44	10.34	10.14	9.83	9.73	10.69 + 0.83
FreeCO_2(mgl)	2.2	2.2	1.87	1.76	1.32	1.21	1.54	1.65	1.76	1.98	2.42	2.42	1.86 ±0.40
pH	7.52	7.56	7.78	7.7	7.8	7.48	7.58	7.66	7.55	7.47	7.48	7.2	7.57 ±_ 0.16
Hardness (mg I^1)	46	60	74	96	78	70	58	58	56	52	38	34	60.00 ±17.48
Chlorides (mg I^1)	5.11	5.3	5.68	5.81	5.12	5.52	5.96	5.39	5.82	5.68	4.26	4.2	5.32 ±0.58
Alkalinity (mgr^{1-1})	45	50	50	52	48	50	60	54	50	46	47	38	49.17 ± 5.31
Phosphates (mg 1')	0.01	0.0	0.054	0.176	0.354	0.02	0.032	0.226	0.0	1.095	0.66	0.402	0.25 ± 0.34

Table 3: Monthly Variations in Physico-chemical Parameters of Bhagirathi River Recorded at the Impacted Site (S2) for the Period September 2004 - August 2005

Parameters	*Sep*	*Oct*	*Nov*	*Dec*	*Jan*	*Feb*	*Mar*	*Apr*	*May*	*Jun*	*Jul*	*Aug*	*X ± S.D.*
Air temperature (°C)	27.5	21.1	15.0	16.2	14.0	18.0	19.0	20.2	25.0	26.5	29.0	29.5	21.75 +5.56
Water temperature (°C)	16.2	14.5	13.0	11.5	10.5	13.5	15.5	20.5	19.2	18.5	17.4	17.0	15.61 ±3.11
Water current (msec^{1})	2.9	2.15	2.12	1.9	1.12	0.99	1.91	2.8	3.1	3.64	4.08	4.1	2.57 ± 1.05
HMD(m)	1.85	1.37	1.03	0.49	0.95	0.73	0.45	1.05	1.36	1.83	2.75	3.05	1.41 +0.83
Turbidity (NTU)	56	55	28	60	49	30	32	98	19	396	628	829	190.0 ± 274.70
Transparency (m)	0.1	0.11	0.18	0.15	0.11	0.15	0.13	0.12	0.14	0.03	0.01	0.0	0.10+0.06
Conductivity (µ Scnr^{1})	0.11	0.15	0.18	0.2	0.19	0.15	0.12	0.1	0.103	0.104	0.08	0.084	0.13 ±0.04
TDS(mgl')	32	26	44	42	30	30	28	50	32	48	1100	90	46.83 ± 26.44
Dissolved Oxygen (mg^{-1})	9.22	9.52	9.72	9.83	9.93	10.74	10.14	10.05	9.73	9.52	9.12	9.02	9.71 ± 0.48
Free CO, (mg^{-1})	3.3	3.3	3.08	2.42	1.98	2.2	2.42	2.42	2.64	3.3	3.74	3.%	2.90 ± 0.64
pH ”	7.75	7.5	7.85	7.85	8.01	7.75	7.55	7.5	7.52	7.44	7.8	7.31	7.65 ± 0.21
Hardness (mg^{-1})	58	64	64	102	100	64	50	58	56	56	42	40	62.83 ± 19.49
Chlorides (mg^{-1})	5.24	5.38	5.94	5.92	5.52	5.94	6.52	5.81	6.24	6.1	5.94	5.82	5.86 ± 0.36
Alkalinity (mg^{-1})	52	55	58	60	54	56	60	62	56	50	49	40	54.33 ± 6.04
Phosphates (mg^{-1})	0.13	0.19	0.208	0.36	0.324	0.252	0.064	0.08	0.008	0.792	0.55	0.836	0.32 ± 0.28

to the benthic macroinvertebrates. Hydrological attributes of the sites SI and S_2 at the river Bhagirathi are presented in Table 2 and Table 3 respectively. Atmospheric temperature was found to be maximum (29.5°C) in August and minimum (14.0°C) in the month of January in the study area. There was no considerable difference in the atmospheric temperature at the reference site and the impacted site. The water temperature was found maximum (20.0°C at S_1 and 20.5°C at S_2) in April and minimum (9.5°C at S and 10.5°C at S_2) in January. Water temperature was not found to follow the trend of air temperature during summer and monsoon months because Bhagirathi river is a snowfed river and thus showing the maximum values in the month of April instead of August. However, the annual mean water temperature was recorded higher (15.61°C) at the impacted site than the reference site (14.75°C). It may be attributed to the dam construction activities which may be responsible for increasing the water temperature at S. Water current was comparatively higher at S (2.57 m sec^{-1}) than S1 (1.87 m sec^{-1}) throughout the year. The water was more turbid at the impacted site (190.00 NTU) than the reference site (139.42 NTU) which may be due to the dumping of stripping and digging materials of tunnels. Overall turbidity was highest during the monsoon months (July-August) due to heavy precipitation in the catchment area. Dissolved oxygen was considerably influenced by the dam construction activities. It was found 9.71 mg l^{-1} at the impacted site compared to the reference site (10.69 mg l^{-1}).

Macroinvertebrate density and diversity

Density composition of macroinvertebrates of Bhagirathi river at the reference site (S1) was represented by the nymphs and adults of the taxa of Ephemeroptera (9), Trichoptera (3), Diptera (4), Coleoptera (5), Hemiptera (2), Plecoptera (2), Mollusca (1) and Odonata (1). However, this diversity composition of macroinvertebrates was altered by the dam construction activities. The diversity composition was represented by the taxa of Ephemeroptera (6), Trichoptera (1), Diptera (2), Hemiptera (1) and Plecoptera (1) at the impacted site (S2). The members of the Coleoptera, Mollusca and Odonata were conspicuously absent throughout the study period at the

Table 4: Monthly Variations in the Density (ind. m^{-2}) of Macroinvertebrates dwelling in the Natural Environment (S1) of Bhagirathi for the Period September 2004 to August 2005

Macroinvertebrates	*Sep*	*Oct*	*Nov*	*Dec*	*Jan*	*Feb*	*Mar*	*Apr*	*May*	*June*	*July*	*Aug*
Ephemeroptera												
Heptagenia	175	175	150	225	375	225	175	125	100	75	25	25
Leptophlebia	25	75	100	125	150	50	–	25	–	–	–	–
Caenis	100	25	50	25	100	125	75	–	50	25	–	–
Cloeon	–	–	25	–	50	75	25	–				
Baetis	25	100	75	50	225	300	125	50	75	50	25	–
Centroptilum	–	–	–	75	100	50	50	25	–	25	25	–
Ephemerella	–	75	–	150	75	125	75	50	25	–	25	–
Ecdynums	–	75	–	50	–	25	–	25	–	–	–	–
Siphlonuridae	–	–	25	100	50	50	25	–	–	–	–	–
Trichoptera												
Hydropsyche	50	100	150	200	150	175	100	75	50	50	25	25
Rhyacophila	25	–	25	50	75	75	50	50	25	–	–	–
Glossosoma	–	–	–	–	–	–	–	–	50	25	–	–
Diptera												
Tendipes	50	50	75	175	200	150	125	50	150	150	75	50
Tabanus	–	50 .	–	75	125	100	25	–	50	25	–	–
Antocha sps.	–	–	–	50	–	25	–	–	–	–	–	–

Psychoda	–	–	–	–	–	–	25	–	–	–	–	–
Coleoptera												
Ochthebius	–	–	–	25	50	75	75	25	25	–	–	–
Dineutussp.	–	–	25	75	100	100	50	50	25	–	–	–
Amphizoa lecontei	–	–	25	–	–	–	50	25	25	25	–	–
Agabinus glabrellus		25	50	–	75	–	25	25	25	–	50	25
Peltodytes	–	–	25	–	–	–	–	–	–	–	–	–
Hemiptera												
Corexia	25	–	50	75	100	50	–	25	–	–	–	–
Hesperocorixa	–	50	25	25	–	–	75	25	–	–	–	–
Plecoptera												
Brachyptera	–	–	25	–	25	–	25	–	25	–	–	–
Acroneuria	–	25	25	75	50	75	25	25				
Odonata												
Hagenius	–	–	–	–	–	–	50	–	–	–	–	–
Mollusca												
Lymnaea	–	–	–	–	–	–	–	25	50	–	25	–
Total	500	775	775	1650	1925	2050	1125	750	725	525	325	200

influenced site (S2). In addition to the qualitative change, a significant quantitative influence in macroinvertebrates in terms of their density was recorded due to dam construction activities. The density of macroinvertebrates was maximum (2050 ind. m^{-2}) during the month of February and minimum (200 ind. m^{-2}) in August at S1, which were found maximum 325 ind. m^{-2} in December and minimum 25 ind. m^{-2} in the monsoon months at S2. The mean density of ephemeropterans was reduced from 468.75 ind. m^{-2} to 110.42 ind. m^{-2} due to dam construction activities. Mean density of trichopterans was also drastically altered. It reduced from 133.33 ind. m^{-2} to 16.67 ind. m^{-2}.

A considerable change in the Shannon-Wiener diversity index (H) of macroinvertebrates was also noticed due to the dam construction activities. It reduced from 3.44 to 1.98 at the influenced site.

Discussion

Most aquatic habitats, particularly free flowing waters with acceptable water quality and substrate conditions, support diverse macroinvertebrate communities in which there is a reasonably balanced distribution of species among the total number of individuals present. Such communities respond to changing habitats and water quality. Under the present study, the substrate composition was found to be degraded at the impacted site (S2) than the reference site (S1). The structure and composition of substrata are one of the main factors controlling the composition of macroinvertebrate assemblages. The bottom substrate composition at S1 (reference site) ranged from small pebbles to big boulders. The macrozoobenthos thrived well in the natural environment (943.75 ind. m^{-2}). However, the density of macroinvertebrates reduced significantly (166.67 ind. m^{-2}) at the impacted site (S2). It may be due to the drastic change in the structure and composition of substrata at this site. Therefore, they are not able to get conducive refuge at the degraded site. The impacted site is silted. It is generally accepted that the sand is a poor substrate for macroinvertebrates, due to its instability and tight packing of sand grains which reduces the trapping of detritus and availability of oxygen. Few macrobenthologists have reported that the distribution of macroinvertebrates was more closely related to substrate of the stream than to water quality[13, 20]. There is a close relationship between substrate type and diversity of zoobenthos. The physically complex substrate types (leaves, gravel, wood and macrophytes) generally support more diversity than structurally simple substrate such as sand and bedrock[21, 22]. In the present study, there is very simple substrate composition at the infected site and having big boulders which may be the cause of poor diversity of macroinvertebrates at this site.

Water flow, temperature and substrates are the major factors determining the composition and abundance of benthic invertebrates. It is not possible to isolate single causal mechanism for changes in diversity, as they are consequence of a chain of complex interactions[23]. Several factors like water temperature, water velocity, riparian vegetation, embankment, turbidity, hydromedian depth, total dissolved solids and transparency were known to influence the distribution of benthos other than substrate composition. Water temperature was found to have a negative correlation ($r = -0.386$ at S1 and $r = -0.419$ at S2) with macroinvertebrate diversity. The same kind of relationship between water temperature and macroinvertebrates was found by other workers[24].

Macroinvertebrate density was found to be maximum (2,050 ind. m^{-2}) in February and minimum (200 ind. m^{-2}) in August in the natural environment of Bhagirathi. Low density during monsoon months (July-August) may be due to the heavy rainfall in the area and flash floods degraded the aquatic environment and structure and components of substrates. The deleterious influence of high or excessive flow of water on aquatic life has also been studied by several workers[25-27]. The diversity of macroinvertebrates was found to have a negative correlation with water current (S1: $r = -0.834$, $p < 0.001$; S2: $r = -0.732$, $p < 0.01$) under the present study. The same result of flow regime on macroinvertebrate communities was also noticed by other workers[28-29].

The diversity of macroinvertebrates was found to be negatively correlated with turbidity (S1: $r = -0.611$, $p < 0.05$; S2: $r = -0.897$, $p < 0.01$), total dissolved solids (S1: $r = -0.557$, $p < 0.05$;

S2: $r = -0.939$, $p < 0.001$) and hydromedian depth (S1:$r = -0.870$, $p < 0.001$; S2: $r = -0.827$, $p < 0.001$). It may be due to the fact that sunlight can not penetrate through the turbid high volume of water and thus hinder the growth of primary producers, which ultimately affect the density of benthos feeding on them. The diversity of macroinvertebrates was found to be highly influenced by dissolved oxygen ($r = 0.834$, $p < 0.001$) at the reference site. However, these were moderately influenced by dissolved oxygen ($r -0.593$, $p < 0.05$) at the impacted site. These observations are in conformity with the inference drawn on the relationship between dissolved oxygen and macroinvertebrates dwelling in running waters[30].

A significant difference in the density ($t = 4.32$; $p < 0.001$) and diversity ($t = 4.71$; $p < 0.0002$) of macroinvertebrates between the sampling sites S1 and S2 was observed. A significant difference was also recorded in the transparency ($t = 2.64$; $p < 0.02$), dissolved oxygen ($t = 3.55$; $p < 0.002$), free CO, ($t = 4.78$; $p < 0.0001$), chlorides ($t = 2.77$; $p < 0.01$) and alkalinity ($t = 2.23$; $p < 0.01$) between the reference site and impacted site. The difference in hydrological parameters and macroinvertebrates density and diversity may be due to the dam construction activities.

Thus, the present study clearly indicates a significant depletion of density and loss of diversity of macroinvertebrates dwelling at Bhagirathi river as a consequence of several dam construction activities of Tehri Dam Project.

Conclusion

Monitoring of the impact of Tehri dam construction on aquatic macroinvertebrates diversity of Bhagirathi revealed that the members of ephemeroptera and trichoptera were drastically influenced due to alteration of the habitat environment. Shannon-Wiener diversity index reduced from 3.44 (reference site) to 1.98 (impacted site).

Acknowledgements

The authors acknowledge the financial assistance given by the Indian Council of Agriculture Research, Govt of India under its major research project (F.No. 4(21)/ 2002-ASR-I). The authors also thank Dr. Deepak Singh for extending help in many ways during the course of study.

References

1. Schnitter, N.J., A history of dams, A.A. Ballkema, Rotterdam, p.1 (1994).
2. Horwath, E., Angel, M., and Municio, T., Impacts of dams on fish fauna/ feasibility of mitigation measures, 2nd Int. PhD symposium, Budapest, 1-8 (1998).
3. Norris, R.H. and Norris, K.R., The need for biological assessment of water quality: Australian perspective, *Australian Journal of Ecology, 2,* 1-6 (1995).
4. Ward, J.V. and Stanford, J.A., Ecological factors controlling stream zoobenthos with emphasis on thermal modification of regulated streams, *Ecology of Regulated Streams,* Plenum Publishing Corporation, pp. 35-53 (1979).
5. Valentin, S., Wasson, J.G. and Philippe, M., Effects of hydropower peaking on epilithon and invertebrate community trophic structure, *Regulated Rivers,* 10, 105-119 (1995).
6. Moog, O., Quantification of daily peak hydropower effects on aquatic fauna and management to minimize environmental impacts, *Regulated Rivers,* 8, 5-14 (1993).
7. Collier, K.J. and Smith, B.J., Sticky trapping of adult mayflies, stoneflies and caddisflies alongside three contrasting streams near Hamilton, New Zealand, *New Zealand Natural Sciences,* 22,1-9 (1995).
8. Quinn, J.M., Copper, A.B., Davies-Colley, R.J., Rutherford, J.C. and Williamson, R.B.-, Land-use effects on habitat, water quality, periphyton and benthic invertebrates in Waikato, New Zealand, Hill country streams, *New Zealand Journal of Marine and Freshwater Research,* 31, 579-598 (1997).
9. Walsh, C.J. Sharp. A.K., Breen, P.F. and Sonneman, J.A., Effects of urbanization on streams of the Melboum region, Victoria, Australia. I. Benthic macroinvertebrate communities, *Freshwater Biology,* 46, 535-551 (2001).

10. Das, S.M., Teaching and Research in Limnology in India, Proc. Symp. UGC NCES, Bangalore, 67-72 (1971).
11. Paul, S. Mitra and Nandi N.C., Impact of anthropogenic activities on Hoogli estuary with special reference to the distribution pattern of intertidal macrozoobenthos, in A.K. Aditya and P. Haldar, Proceedings of the National Seminar on Environmental Biology, pp. 94-101 (1998).
12. Bhatt, J.P., Jain, Ajay, Bhaskar, A. and Pandit, M.K., Preimpoundment study of biotic communities of Kistobazar Nala in Purulia, West Bengal, *Current Science,* 81(10), 1332-1337 (2001).
13. Sharma, R.C., Bhanot, G. and Singh, Deepak, Aquatic macroinvertebrate diversity in Nanda Devi Biosphere Reserve, India, *The Environmentalist,* 24, 211-221 (2004).
14. Sharma, R.C., Water quality, composition and abundance of benthos in the Bhagirathi river, Garhwal Himalaya, *Indian J.Ecol,* 13(1), 133-137 (1986).
15. Sharma, R.C., D.Sc. Thesis, H.N.B. Garhwal University, pp. 118(1991).
16. APHA, Standard methods for the examination of water and waste water, American Public Health Association, New York, USA, (20th Ed.) (1998).
17. Bovee, K.D. and Milhous, R., Hydraulic simulations in instream flow studies: theory and techniques, United States Fish and Wildlife Services, Cooperative in stream flow group, in stream flow information paper, pp. 21 (1978).
18. Ward, H.B. and Whipple, G.C., Freshwater Biology, John Wiley and Sons (2nd Edn.) New York, pp. 1248 (1992).
19. Shannon, C.E. and Wiener, W., The mathematical theory of communication, University of Illinois Press, Urbana, USA (1964).
20. Ruggiero, M. A. and Merchant, H.C., Water quality, substrate and distribution of macroinvertebrates in the Patuxent river, Maryland, *Hydrobiologia,* 65 (2), 183-189 (1979).
21. Angradi, T.R., Inter-habitat variation in benthic community structure, function and organic matter storage in 3 Appalachian headwater streams, *J. Am. Benthos L. Soc. (USA),* 15, 42-63 (1996).

22. Hawkins, C.P., Substrate association and longitudinal distribution in species of Ephe-merellidae (Ephemeroptera: Insect) from Western Oregon, *Freshwater Invertebrate Biology,* 3(94), 18-88 (1984).
23. Boon, P.J., The impact of river regulation on invertebrate communities in U.K., *Regulated Rivers, 2,* 389-409 (1988).
24. Barrett, P.H., Potassium concentration in fertilized trout lakes, *Limnol Oceanogr,* 26, 510-520 (1957).
25. Stevenson, R.J., in R.J., Bothwell, M.L. and Lowe R.L. (Eds), The stimulation and drag of current, in R.J. Stevenson, M.L. Bothwell and R.L. Lowe, (eds.), Algal Ecology: Freshwater Benthic Ecosystem, Academic Press, pp. 125-156 (1996).
26. Cushman, R.M., Review of ecological effects of rapidly varying flows downstream from hydroelectric facilities, *North American Journal of Fisheries Management,* 5, 330-339 (1985).
27. Biggs, B.J.F., Stevenson, R.J. and Lowe, R.L., A habitat matrix conceptual model for stream periphyton, *Fresh Hydrobiol,* 143(1), 21-56 (1998).
28. Erman, D.C. and Mahoney, S.D., Calif, Recovery after logging in streams with and without buffer stripes, Water Res. Center, University of Calif., Northern California, pp. 50 (1983).
29. Kalmers, E., Distribution of Plecoptera and Ephemeroptera in relation to altitude above sea level and current speed in mountain waters, *Pol. Arch. Hydrobiology,* 14 (27), 29-42 (1967).
30. Bretschko, G. and Klemens, Quantification method and aspect in the study of the interstitial fauna of the running waters, WE., *Stygologia,* 2, 297-316 (1986).

14

Impact of Tehri Dam on Aquatic Biodiversity[1]

Anil Gautam

Editor's Introduction: This study has been undertaken by an NGO People's Science Institute for the Uttarakhand Environment Protection and Pollution Control Board. The results indicate that the biodiversity of aquatic organism is drastically impacted by the Tehri Dam.

Applying physico-chemical tests is the most widely followed method for monitoring the quality of water bodies. To supplement physico-chemical techniques of water quality assessment, bio-monitoring has emerged as an important tool in understanding the overall health of an aquatic ecosystem and to ascertain the cumulative impact of the pollutants on it. Bio-monitoring—using benthic macro-invertebrates—as as indicator organisms, is a cost-effective method of water quality assessment and is now being widely used throughout the world. Benthic macro-invertebrates spend most of their early life in the substratum of freshwater bodies. The adults, however in many cases, have

1. Adapted from "Bio-monitoring of the Ganga river and its tributaries in Uttaranchal," People's Science Institute, Dehradun, 2005.

wings and fly around in search of suitable freshwater environment for laying eggs.

The Central Pollution Control Board (CPCB) has established and validated bio-monitoring (using benthic macro-invertebrates as indicators) as a cost-effective system for water quality assessment and initiated nationwide water quality assessment exercises based on this technique. The CPCB has also undertaken an active program of bio-monitoring in problem areas.

This study covers eight important rivers, namely-Alaknanda, Dhauliganga, Nandakini, Pindar, Mandakini, Bhagirathi, Bhilangana and the river Ganga (from Dev Prayag to Haridwar). Major habitations along these rivers are Badrinath (represents high influx of devotees from April to September), Joshimath, Chamoli, Nandprayag, Karanprayag, Gauchar, Rudraprayag, Srinagar, Uttarkashi, Rishikesh and Haridwar. Most of these habitations have religious importance, and receive a heavy burden of tourists in different parts of the year. The present report presents the first phase of sampling was carried out from mid-November to mid-December 2003.

Methodology

Unfortunately, organisms are not randomly distributed over a water-body and different habitats (sand, mud, gravel or organic material) and many of them are constantly moving. Therefore, it was desirable to keep sampling time constant (1hr) at every sampling station selected for the study.

In several sections the substratum was rocky or gravel bed area hence stones were picked from the river body and rubbed to collect organisms from the stone surface. The collected organisms were kept in the collection tray. Each rubbed stone was placed outside of the sampling area.

Sediment was collected with the help of a hopper/shovel and poured through an appropriate sieve using additional water as per need. Mud or very fine sand will pass through the sieve leaving the organisms behind. The animals were sorted carefully and placed it in the collection tray using forceps.

For sampling the organisms, which are found in sediment, sediment was collected directly in sieve or filter net.

Some of the collected samples were identified on the spot with the help of identification key. Others were preserved in 4 per cent formalin and were carried to the laboratory for identification.

Benthic Macro Invertebrates

Families such as Baetidae, Leptoceridae, Pothaminthidae, Heptageniidae, Goeridae, Siphonuridae, Nemouridae and Ephermelidae were found almost all over the river from Gangotri to Haridwar.

Lepidostomatidae and Polycentropodidae families were found mostly in the upper stretch of river Bhagirathi.

Psephinidae, Hydropsychidae, Perlidae, Perlodidae and Ephemeridae were found from middle stretch to lower stretch, i.e. in Alaknanda, Nandakini Pindar, Mandakini and ultimately in Ganga.

Limnocentropodidae, Gyrinidae, Vellidae, Culi-cidae, Uenoididae and Rhycophilidae were the families found mostly in the middle stretch, i.e. from Chamoli downstream to Rudraprayag downstream.

Chironomidae, Oligocheata, Gomphidae, Hygro-bidae, Simulidae, Lestidae and Tabanidae were the families found only in the lower stretch indicating organic pollution loads contributed by the major towns such as Srinagar, Devprayag, Rishikesh and Haridwar.

In both Bhagirathi and Alaknanda, the number of benthic families and individuals found in the middle and lower stretches was greater than those found in the upper stretch. Further, the number of benthic families and individuals in these two major rivers was less than those found in their tributaries. This may be due to the low discharge, velocity and depth of the tributaries. Another reason may be the difference in the composition of the substratum (habitation) of the main river and the tributary. Moreover, some families, which are pollution sensitive, were already present in the water. Also the addition, in the middle stream, of the discharge (organic matter) from the townships led to an increase in the number of pollution loving species. Since there was no crisis of dissolved oxygen throughout the studied stretches, the pollution sensitive species which should have disappeared due to the increase in organic pollution load, were also present in the middle stretch. This may be a possible reason for the large number of families and individuals present in the middle stretch.

Impact of Tehri Dam

At two sampling points on the Bhagirathi (GBH06 and GBH07), the benthic individuals were completely absent as shown in Table 1. A possible reason for this may be the construction of the dam at Tehri, which was upstream to the sampling locations. At Tehri, the river Bhagirathi was canalized and was diverted from its normal course through two tunnels during the course of which it picks up a lot of silt and sand. This led to a drop in the transparency of the river and possibly a disturbance of the substratum (habitation). This may have led to a disappearance of the benthic individuals at these locations.

Table 1: Biodiversity of River Ganga Upstream and Downstream Tehri Dam

Sl. No.	*Station Code*	*Biodiversity individuals*	*Number of of families*	*Number*
4.	GBH 04 u/s	3.29	48	13
5.	GBH 05 u/s	3.18	50	12
7.	GBH 06 d/s	0.00	0	0
26.	GBH 07 d/s	0.00	0	0

Similar reduction in the Biological Monitoring Working Party Score (BMWP) and Sequential Comparison Index (SCI) is seen as shown at Table 2.

Table 2: BMWP and SCI of River Ganga and its Tributaries

Sl. No.	*Station Code*	*BMWP*	*SCI*
4.	GBH 04	8.2	0.40
5.	GBH 05	8.2	0.67
7.	GBH 06	0.0	0.00
26.	GBH 07	0.0	0.00

BMWP – Biological Monitoring Working Party Score.
SCI – Sequential Comparison Index.
BWQC – Biological Water Quality Criteria.

Conclusion

The water quality of the River Bhagirathi up to Uttarkashi was almost the same. In downstream of Uttarkashi, the water quality slightly degraded. This is indicated by the physico-chemical parameters and the biomonitoring results. The river water quality recovered up to upstream of Tehri. Here after, the water quality became very poor due to the Tehri dam construction.

Upstream of Srinagar, the water quality of the River Alaknanda was almost constant. It started to degrade after Srinagar and the situation remained the same up to the up stream Rishikesh.

It further deteriorated in Rishikesh due to the Chilla diversion barrage at Rishikesh,[1] where the water was accumulated. At the subsequent station, the bulk of the water from the River Ganga is diverted to the Chilla hydel plant and the Ganga there after is predominantly composed of the rivers Song and Suswa. Hence the water quality of the Ganga was maintained.

At the Bhimghoda diversion barrage in Haridwar, the East and the West Ganga Canals were diverted and very little water remains in the main stream. The effluent from the sewage treatment plant at Jagjitpur and the township waste is also discharged into the main stream. Hence the water quality at

1. Mainly in increase of BOD.

this location was the worst observed anywhere through out, the studied stream stretches.

Except for two sampling locations downstream of Tehri Dam on River Bhagirathi (GBH06 and GBH07), the water quality at sampling locations in the middle and down stream of the River Alkananda is inferior to that of the River Bhagirathi. This may be due to more habitation in the Alaknanda catchment area as compared to the Bhagirathi catchment.

PART 5: CONCLUSION: ALTERNATIVE WAYS OF GENERATING HYDROPOWER

Introduction

The previous chapter have established that construction of dams leads to the deterioration of subtle, chemical and biotic qualities of water. This sector addresses the question: What next?

Hydropower proponents sometimes accept the negative environmental impacts. However, they hold that this environmental cost has to be borne in order to secure economic growth. First chapter in his sector provides are estimate of monetary loss to people of the country from the deterioration of water quality. It is shown that this loss is much greater than benefits from electricity generation. In other words, we are incurring both economic and environmental loss from Tehri Dam.

The second paper suggests that part of the river water can be diverted for generation of hydropower and remaining part may be allowed to flow free. This latter part with maintain quality of river water and the loss to environment and economy from construction of dams can be prevented.

15

Monetary Value of Quality of Water of River Ganga[1]

Bharat Jhunjhunwala

Editor's Introduction: An assumption is generally made that no loss is incurred by the society by damming the rivers. This study shows that such is not the case. The monetary loss to people from destruction of the subtle qualities of waters in the dams is much greater than the monetary benefits from generation of hydropower.

Uttarakhand Government is making a cascade of Hydro-electric Projects (HEPs) on River Ganga. These dams are going to affect the quality of water at Devprayag, Rishikesh, Haridwar and downstream at Prayag, Kashi and Ganga Sagar. HEPs claim that damming the flow of water will provide benefits from generation of power to the people and will *not* reduce the spiritual power of the water at these places. Tehri Hydropower Development Corporation (THDC), which has built the Tehri Dam, insists that there has been no deterioration of quality of water downstream from Tehri Dam. This study is made to assess the impact of Tehri dam on the spiritual impact of taking bath at Devprayag, Rishikesh and Haridwar as assessed by

1. Adapted from *Economics of Hydropower,* Kalpaz Publications, Delhi, 2008.

pilgrims. An attempt is made to assess the monetary value of benefits accruing to the people from free flow of the river.

Methodology

The objective of the study is to assess whether and to what extent the construction of Tehri dam has affected the spiritual power of River Ganga. A questionnaire was made to ask respondents about impact of dam on various aspects of worship. Naresh Chandra Puri, Diploma Pharma, resident of Village Lakshmoli, Dt Tehri, went to these pilgrim centres in December 2007-January 2008 and sought responses from the pilgrims coming to take bath or to collect water for taking home.

No sampling was done. Assistance of any local person was not sought lest the mediator influence the results. The enumerator asked people he met spontaneously about the impacts.

The income of respondents is given at Tables 1 and 2.

Table 1: Income of Pilgrims Coming from Outside

Sl. No.	*Location*	*Average*	*Median*
1.	Devprayag, n=27	15,355	10,000
2.	Rishikesh, n=37	8,166	5,640
3.	Haridwar, n=61	8,473	5,000
4.	Average, n=125	10,665	6,880

Table 2: Income of Local Pilgrims

Sl. No.	*Location*	*Average*	*Median*
1.	Devprayag, n=21	3,042	2,000
2.	Rishikesh, n=21	11,623	6,000
3.	Haridwar, n=15	4,720	3,000
4.	Average, n=57	6,462	3,667

Nature of Benefit obtained from River Ganga

The questions asked and replies received have been given in tables below.

Question: What is the benefit you have obtained from the blessings of River Ganga?

Response: Seventy seven per cent pilgrims obtained mental peace while 26 per cent obtained health benefits. The Ganga waters appear to have the power to relax the mind and body of the pilgrim. This confirms that water of Ganga indeed carries subtle psychic charges. Details are given at Table 3.

Table 3: Per cent of Pilgrims deriving a particular type of benefit (Per cent of Respondents replying in affirmative).[1]

Sl. No.	*Type of Benefit*	*Unit*	*Devprayag, n=55*	*Rishikesh, n=93*	*Haridwar, n=94*	*Combined, n=242*
1.	Mental Peace	Per cent	75	71	84	77
2.	Health	Per cent	25	19	33	26
3.	Business	Per cent	18	5	19	14
4.	Child	Per cent	18	9	8	12
5.	Service	Per cent	5	9	23	12
6.	Success in Examination	Per cent	11	9	8	9
7.	Others	Per cent	31	13	17	20

Question: How much value do you place on the benefit obtained?

Response: In response to this question a large number of pilgrims replied in terms of 'invaluable', 'lacs' or 'priceless'. These have been mentioned in Table 4.

Table 4: Value of Benefit obtained

Sl No	*Parameter*	*Devprayag, n=11*	*Rishikesh, n=6*	*Haridwar, n=9*	*Average, n=26*
1.	Average	29,427	59,041	66,177	51,548
2.	Median	1,000	26,500	10,000	12,500
3.	Number responding in priceless, invaluable, etc	25	23	4	52

1. The total can exceed 100 per cent because many pilgrims have derived more than one type of benefit.

The average or median values are based on a small number of respondents, 26. Nearly double this numbers have said that the benefit derived is priceless etc. In view of this, we take the higher average value of ₹ 51,548 as the value of benefit derived by pilgrims, and not the lower median value.

Decline in Quality of Water due to Construction of Dams

Question: What is the impact of Tehri Dam on quality of waters of Ganga?

Response: The responses are given at table 5. The decline in quality of water is less as we move away from Tehri. This is likely because the water gets a chance to regenerate some qualities like Dissolved Oxygen. Further, there is higher flow of water in the winters and summers due to construction of Tehri Dam. This may be a positive aspect that was noted by Pilgrims.

Table 5: Impact of Tehri Dam on Quality of Water (Percent Pilgrims Responding in Affirmative)

Sl. No.	*Type of Pilgrim*	*Was Better Previously*	*Is Better Now*	*No Difference*	*Net, Per cent Reporting Previously Better*
1.	Devprayag, n = 41	88	12	–	76
2.	Rishikesh, n = 87	47	15	38	32
3.	Haridwar, n = 78	38	26	36	12

Benefit to Pilgrims from Removal of Tehri Dam

In order to assess the impact on pilgrims due to change in quality of waters the pilgrims were asked to give a per cent value to the increase or decrease in the impact due to change in quality of water. These are given at table 6.

Table 6: Change in impact due to the removal of Tehri Dam, Per cent, Average

Sl. No.	*Location*	*Pilgrims Reporting Increase*	*Pilgrims Reporting Decrease*
1.	Dev Prayag, n=31, 3	46.3	43.3
2.	Rishikesh, n=52, 11	46.0	30.0
3.	Haridwar, n=42, 17	29.5	42.7
4.	Average, n=125, 31	40.6	38.7

Estimate of benefits to the Pilgrims from Removal of Tehri Dam

Last line of Table 6 tells us that 125 respondents reported increase of benefits by 40.6 per cent if Tehri is removed. On the other hand 31 respondents reported a decline of benefits by 38.7 per cent if Tehri Dam is removed.

We can use these figures to make an estimate of effect of removal of Tehri Dam on pilgrims.

The weighted average of benefits is calculated as follows:

Benefit to pilgrims (based on last line of Table 6)

= ((Per cent increase × numbers reporting increase) −(Per cent decrease × numbers reporting increase))/ (Total numbers)

= ((40.6 × 125) – (38.7 × 31))/156

= 24.8%

The benefit to pilgrims is increased by 24.8% upon removal of the Tehri Dam.

Present benefit per pilgrim: ₹ 51,548 (Row 1 of Table 4 above).

Increase in benefit per pilgrim from removal of Tehri Dam: 24.8% of ₹ 51,448 = ₹ 12,784.

The number of pilgrims is estimated as follows: Dev Prayag 200, Rishikesh 1000, Haridwar 2500, other places before Haridwar 1,300, Total 5,000 per day.

Gain to pilgrims from removal of Tehri Dam: 5,000 pilgrims × 365 days × ₹ 12,784 = ₹ 2333 crore per year.

Further gains would accrue from pilgrims taking waters to their homes. The power of that water will also be reduced. Further benefits will accrue to other non-pilgrim local people residing on the banks of Ganga and non-pilgrim travellers in the area. This is, therefore, an underestimate.

A similar amount may be assessed for pilgrims downstream from Haridwar at Prayag, Kashi, Ganga Sagar, etc. The total loss to the pilgrims from Tehri, therefore, may be assessed at double this amount at ₹ 4,666 crores per year.

Conclusion is that removal of Tehri Dam will lead to benefits of ₹ 4666 crores per year to the pilgrims from improvement in the quality of water.

Generation of electricity from Tehri Dam is about 2700 million units per year. Taking a benefit of ₹ 1.72 per unit to the user[1] the total benefit from electricity from Tehri Dam is estimated at ₹ 4644 million or ₹ 464 crores.

1. The Energy Research Institute (TERI) has estimated that industrial customers' willingness to pay (WTP) is about ₹ 6 per kWh—in constant 2006 prices—in Haryana and Karnataka. We may consider the same value for domestic and commercial consumers. Farmers' WTP for electric power is ₹ 1.8 per kWh in Haryana, ₹ 4.4 per kWh in Karnataka and ₹ 3.5 per kWh in Rajasthan. Average WTP by farmers is ₹ 3.2 per kWh. These translate into ₹ 6.9 for industrial consumers and ₹ 3.7 for farmers (Consumer Price Index was 198 in 2006 and 226 in 2012). Taking tariff at ₹ 5 per kWh for industrial, commercial and domestic consumers the net benefit to consumers will be ₹ 1.9 per kWh (WTP 6.9 – Purchase Price 5.0). Taking tariff at ₹ 2 per kWh for agricultural consumers the net benefit to farmers will be ₹ 1.2 per kWh (WTP 3.2 – Purchase Price 2.0). Taking 75% consumption by industrial consumers and 25 percent consumption by farmers, the weighted average of net benefit to all consumers is estimated at ₹ 1.72 per kWh. Figures for WTP have been taken from *Project Appraisal Document on a Proposed Loan in the Amount of US$ 648 Million to THDC India Limited With The Guarantee of the Republic of India for the Vishnugad Pipalkoti Hydro Electric Project*, Report No: 50298-In, World Bank, June 10, 2011.

Against this the monetary benefits to be obtained by people of the country from removal of Tehri Dam is estimated at ₹ 4666 crores per year—nearly ten times the benefit from use of electricity.

Actually both these estimates are incomplete and inadequate. Additional benefits are obtained due to use of water in irrigation and drinking water. Additional costs are incurred due to damage to environment such as loss of aquatic biodiversity, carbon emissions, loss of forests, etc.

Nevertheless, I have given this estimate to show that the monetary value of benefits from preserving the subtle qualities of waters is much greater than the benefits from electricity.

16

Generation of Hydropower while Maintaining Free Flow

Paritosh Tyagi and Bharat Jhunjhunwala

Editor's Introduction: Previous sections of this book have detailed the spiritual powers of the Ganga. However, I recognise that electricity is also required. In this paper, the authors suggest a way to generate hydropower from the Ganga while conserving her spiritual powers. Generation of hydropower will be less but her invaluable spiritual powers will be conserved.

The current practice is to make a dam or a barrage across the river, create a reservoir, divert the water from the reservoir through a tunnel, pipeline or canal towards the downstream side of the river, utilise the head of water for generating hydropower in a powerhouse fitted with turbines and finally restore the flow in the river through the tail race.

Another current practice is to make a large reservoir (typically 20 to 40 km long); and water is drawn from top of the reservoir as in the under construction Srinagar HEP on Alaknanda.

Generally, the tunnel-based alternative is used where volume is less and gradient is high (as at Vishnu Prayag on

Alaknanda) while reservoir-based alternative is used where the volume is large and gradient is less (as in Srinagar on Alaknanda).

The most important effect of the existing practice is on the nature of the river due to impoundment and diversion of flow. The sediments carried by the river are arrested; the aeration of water almost ceases; the biota is severely disturbed; migration of fish is obstructed; aquatic biodiversity is badly affected; livelihood of people dependent on the river is threatened; pilgrims and tourists are disappointed by the deterioration in the condition of the river; land subsidence takes place; depletion takes place in the yield of local springs and damage to dwellings takes place during blasting for making the tunnel.

In place of the above pattern the alternative of Off Stream Storage may be considered for generating hydropower.

Storage is essential for generating peaking hydropower. Currently, the reservoir upstream of the dam provides storage of water which is released during peak hours. However, the water can also be stored in off-stream tanks somewhat like a large sedimentation tank. Water can be fed into such tanks by diverting part of the water from the river by providing openings at the bends to throw out water due to centrifugal force. As a less-preferred alternative, an obstruction or spur can be made over part of the width of the river to divert part of the water without blocking it completely. Such an obstruction had been made at Bhimgoda in Haridwar under an agreement arrived at between representatives of the people and the British Government in 1916 although this had not been made for hydropower. We refer to Bhimgoda only to draw attention to the fact that such partial obstructions are technically feasible for hydropower as well. These alternatives will enable free flow of the river and maintain the pristine natural quality of water.

Storage of water in off-stream tank will entail some loss of gradient as well as increase in project cost.

The discharge from the upstream project can be fed into the intake of the downstream project and the discharge from the last project can put in a canal and used for irrigation. That will ensure that the pristine quality of the remaining free-flowing water is not disturbed throughout the length of the river (Fig. 1).

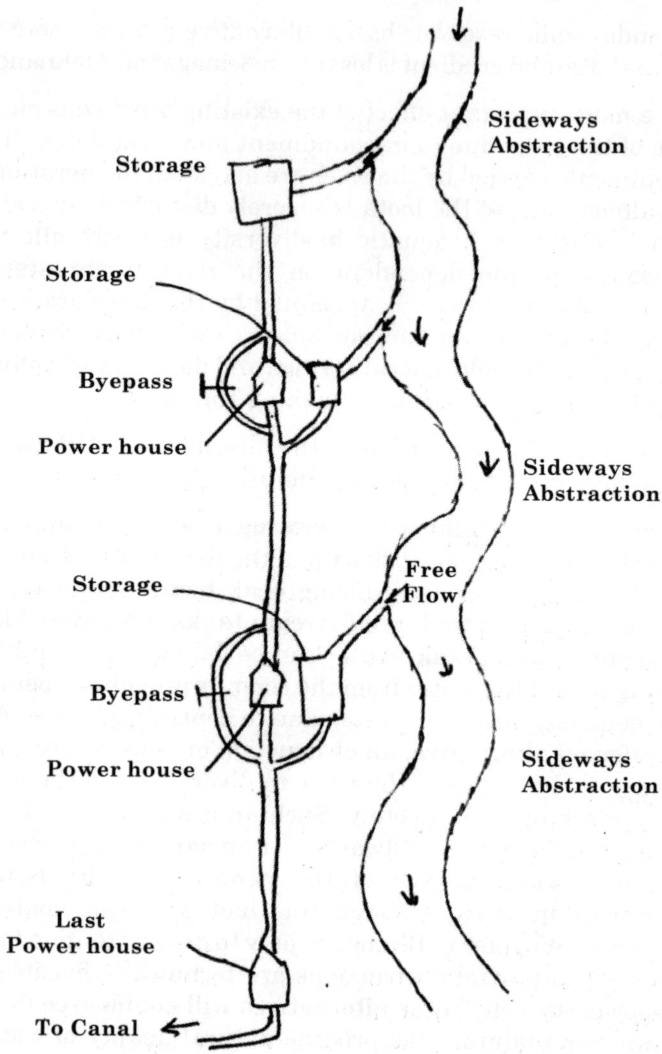

Fig. 1: Schematic Representation of Alternative Method of Generation of Hydropower.

Off-stream storage will need intensive survey of local site conditions and topography. Only part of the flow will be abstracted by an opening on the side to be pumped or conveyed by gravity to a suitable site for storage. Though this alternative appears reasonable, not much is known about its application. Likely sites for storage may present problems similar to those encountered in locating a reservoir.

A major advantage for off-stream storage projects will be that settled sediment in the off-stream storage can be removed pre-monsoon by discharging all stored water. This removes the problem of silting and silt removal in on-stream storage dams like Tehri. Desilting cannot be done of on-stream storage reservoirs because of continuous inflow of river water.

A near example of sideways abstraction is the Tazimina hydroelectric project in Alaska. Water intake of the project is located at the side of a natural pool that forms upstream a waterfall. A line of sill blocks are placed in the river bed to divert water to the intake. These are low height blocks that do not obstruct migration of fish or downstream flow of sediments. There is no dam. The outlet after generation of hydropower is located at the foot of the waterfall. The project is small—824 kw. However, there are large numbers of sites on Ganga where such natural pools are available and water could be abstracted from here without making a dam (Fig. 2).

Fig. 2: Taziminia Hydropower Project, Alaska, USA. Source: alaskarenewableenergy.org/alaskas-resources types.../ hydroelectric/

Our most preferred option is to abstract water without making any structure on the river bed. A less preferred option is to make a partial obstruction. Herman Wanningen of Netherlands has provided us with a diagram which explains how this can work (Fig. 3).

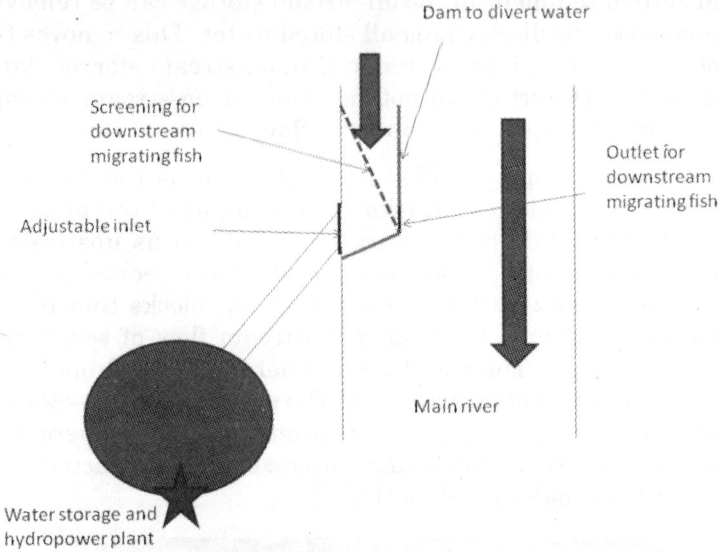

Fig. 3. Schematic Representation of Abstraction of River Water by Partial Obstruction.[1]

Mr Jim MacBroom is a Civil and Environmental Engineer based in Connecticut, USA. He has worked on the development of minimum in-stream flow standards for the State of Connecticut. He currently serves on the American Society of Civil Engineer's National River Restoration and Sedimentation Committees. In a personal communication he writes:

> There are examples of partial diversions without a dam, but you need is a waterfall or steep rapids to provide the head for power generation. (The) Bulls

1. *Courtesy*: Wanningen Water Consult, Netherlands.

> Bridge and Falls Village power stations on the Housatonic... use low weirs at the crest of cascades to divert partial flow to a power station head race canal while the main (aka bypass reach) channel remains wet. *They could have been made without the low weirs and a deeper head race, but would have less available head and develop less power.*

The key question is not technical feasibility. The key question is that making a weir or dam provides increased head at low cost. Abstraction of water without making a weir or dam will reduce the head and correspondingly increase the cost. We must apply our mind to whether the increased cost is justified in view of the social and cultural benefits.

There is no denying the fact that more means of generating power are needed to meet the growing requirement of power and that hydropower is one alternative for this purpose. However, we feel the costs from no-or reduced generation from the rivers are small in comparison to the huge socio-cultural benefits.

Concerned experts in hydrology, land morphology, geology, sociology, environmental science and economics may be encouraged to examine the subject in detail with particular reference to the upper reaches of Himalayan Rivers.

Storage is essential in our country because rain falls during a limited part of the year while power is needed throughout the year. Since generation of hydropower can be easily started and stopped by controlling the flow of water to the powerhouse, it is expedient to generate hydropower to meet the peak demand for power. Need for storage of water is, thus, unquestionable. But it is unfortunate that it is taken as an axiom that river has to be blocked to store water. That storage of water is possible away from the stream deserves serious consideration to avoid the damage to water quality and to the river as a living ecosystem. The following alternatives are suggested in this regard:

1. Water can be stored in off-stream gorges and bowl shaped formations. Such storages can usually be filled with water by intercepting the flow of water from the catchment towards the river. If needed, water can be fed

into such storages by diverting part of the flow of the river at convex bends with the help of centrifugal force of river flow.

2. There is a large volume of storage available for storing water under ground in the Indo-Gangetic plains owing to considerable depletion of groundwater table. By recharging groundwater, two advantages will be gained, viz., irrigation of agricultural fields and increase in base flow of the river.

The aforesaid alternatives will enable free flow of the river and maintain the pristine natural functions of the river and the quality of its water.

PART 6: MORE EVIDENCE OF SUBTLE QUALITIES OF WATER

Introduction

There is growing body of scientific literature on the subtle qualities of water in general, i.e., not specifically related to the impact of dams. Importance of this literature lies in the fact that water must be seen more than merely physical and faceless H_2O.

Hydropower projects are made on the assumption that there is no 'higher' or 'subtle' qualities of water. Therefore, it is contended that generation of hydropower is a win-win circumstance where we get something without losing anything. The energy contained in water flowing from higher to lower elevation is captured without making any harm to the water.

The seven papers in this section provide a glimpse of the evidence that water is more than H_2O. It has capability of carrying thought and memories. Water flowing down the Ganga, therefore, *may* be carrying some special psychic information which may be lost while water passes through hydropower turbines. Indeed, there is some evidence presented earlier in this book that these qualities are impacted by hydropower dams. I concede that these evidences are not as robust as I would like it to be. However, the 'Precautionary Principle' enshrined in the law of the country requires us to err in the side of caution in face of incomplete knowledge.

I present these papers tc show that there is adequate evidence of subtle qualities of water and it is necessary for us to commission more studies to verify these.

17

Implications of Memory of Water for Spiritual and Religious Beliefs[1]

Steve Gamble

Editor's Introduction: This article explains that the capacity of plants to respond to particular persons is due to the water retained in its leaves. Memory of a person who harms or helps the plant is stored in the water contained in the leaves. Accordingly plants respond positively or negatively to different persons.

The author further states that memory of one's own past life or of other past events is stored in water and is retrieved at current period from water.

To appreciate the role the 'memory' of water plays in the scheme we call life, we first need to understand how all living things interact on an energy level, through the medium of water.

In the book–*The Secret Life of Plants*, Peter Tompkins and Christopher Bird chart some very interesting scientific developments relating to the consciousness of plants and the energy interaction and influence between all living things. It is an exciting discovery of how energies interplay with each other and which can directly or indirectly influence other living systems. Discoveries which were so profound by their nature, naturally aroused the interest of the intelligence agencies and the military in the 1950s and 1960s in what may well have been the springboard for their respective present-day non-lethal weaponry—a unique way of influencing and controlling populations of people through radio and microwave radiation without causing a fuss or drawing untoward attention to their actions.

Tompkins and Bird start by looking at the work of Clee Backster, one of America's foremost lie detector examiners. Backster soon discovered by playing with his galvanometer and attaching electrodes to the leaves of plants, he could witness a response in plants to his own thoughts. Galvanised and excited by his findings, Backster started an intensive programme of research into the extraordinary perception of plants. He discovered that plants reacted not only to threats from humans, but also from other animals. Backster's investigations continued to discover if a plant had memory. He set up an experiment where six of his students, blindfolded, drew lots from a hat. On one piece of paper only were the instructions to actually damage the plant. Two plants were used; one was duly damaged by the student acting on the instructions on the piece of paper and the other plant acted as a witness to the event. Backster showed that the surviving plant gave no reaction to five of the six students, but whenever the actual culprit approached it, it reacted and caused the meter to go wild. Backster had effectively demonstrated that plants had memory.

Backster went on to extend his field of investigation into mould cultures, scrapings from the human mouth, blood cells, and even sperm. The results were just as interesting as those he discovered with plants, and sperm cells were quite uncanny in the way they seemed capable of identifying their own donor. After further experiments his findings were published in 1968 in the International Journal of Parapsychology. Backster's work was providing a foundation of belief that there was some form of cellular consciousness common to all living things.

Other scientists followed up on Backster's work with their own experiments and one, Sauvin, found he could control his model train set or even start his car engine by transferring his thoughts to a plant while the plant was connected to the device in question. Sauvin went on to show that this phenomenon could be repeated by controlling the flight of model aeroplanes. By transmitting thoughts to a plant Sauvin could start, stop or alter the speed of a model plane in flight.

As more and more of these scientific experiments and their startling results were written up in journals all over the world, more scientists became involved in similar research programmes. Vogel was one and reasoned that there was a 'psychic energy' at play and that this energy was storable. Vogel noticed that leaves with a large electrical resistance were especially difficult to work with, though fleshy leaves with high water content were best. He concluded that man can and does communicate with plant life and that plants are living objects, sensitive and rooted in space. They radiate energy, which one can feel as they feed this energy into our own force field which in turn feeds back energy to the plant.

In other parts of the world similar experiments were being undertaken which led to developing the use of radio frequencies to stimulate the growth of plants. In 1970, L. George Lawrence read that in the Ukraine, radio frequencies and ultrasonic vibrations had been used to stimulate cereal grain seeds to produce higher yields as far back as the 1930's. In 1970, *Pravda* published an article by professor Ivan Issidorovitch Gunar, head of the Academy's Department of Plant Physiology, which confirmed the presence of electrical impulses in plants, similar to nerve impulses in humans. This report described similar findings which proved there was communication between plants and humans, and a plant's reaction to human intent–be it love or hate–through calmness or fear.

There have to date been numerous experiments and tests which demonstrate clearly how plants and vegetables do react to pain, emotion and love. Experiments continued in many areas of the world on the effects of not only thought/emotion upon the consciousness of plants, but the effect of other vibrations and sounds such as music and ultrasonics.

Tompkins and Bird go on to describe an eight week experiment carried out on summer squashes by broadcasting music to them in their respective chambers from two Denver radio stations. One station specialised in heavy rock and the other in classical music. Those squashes exposed to Hayden et al., grew towards the transistor radio, one of them twisting itself lovingly around it. The other squashes, subjected to the rock music all grew away from the radio and even tried to climb the walls of their glass cage. Similar tests were carried out on other plants and vegetables with the same results. Rock music proved to be detrimental to the well being of living matter, whilst classical music proved to be beneficial.

This is how energies work and every living thing, be it plant, vegetable, animal or human, is composed of the very same energy–Spirit, God, Oneness, Source or whatever label we would like to attach to it–and it is only the way this energy animates the biochemical aspect of this existence that makes living entities different in appearance. We are all one with nature and nature is all one with us. However, due to the Fall of Humans, both in consciousness (soul) and awareness, we no longer have that conception of the overall we, the belonging to and *being* a part of the *collective WE*. We now look at life from the singular aspect of the ego, the *ego of I*, the I AM PRESENCE, and we view life as being composed of individual and separate entities, ourselves included, and so we do not see or feel the plant or animal as an extension of ourselves, or as being one with us. Instead through ego, we now see all life outside of us as being separate, and therefore alien to us. All consciousness is one, we are everything and everything is one. Consciousness just IS...[It is only the intellect and the ego that sees itself as being separate from, and even superior to, the rest of nature, feeling the need for some form of supernatural being to come and save us from our belief-induced-sins.]

The human body is composed of over three-gallons of water and so we humans are highly emotional, and therefore, very receptive to other external influences and energies! The Earth and its atmosphere is made up of around 70-80 per cent water too, and so the Earth (its consciousness) is one huge 'memory bank' of stored information and emotion from the beginnings of time. In this respect the Earth is the Akashic Records and we can tap into this 'memory' at any time because it is all around us. The Universe too is replete with water which is how we can extend our consciousness (soul/awareness) anywhere we choose to – as some people do when remote viewing a distant entity for example.

Our thoughts and beliefs affect the consciousness/energy of everything around us because almost everything is replete with water. For example, we may go to the finest organic restaurant in town, but if the Chefs and their staff are in a bad mood, then that 'emotion/energy' will be passed into the food that they prepare. We, by consuming the food, then 'absorb' that energy/memory/emotion, or can become aware of it. This applies to all processed food that we find available today. There is no escaping this.

Because water stores energy/emotion/information from ALL sources, therefore our emotions, thoughts, beliefs etc., ARE automatically stored externally in the watery environment (consciousness) of the Earth and beyond!

The 'memory' of water – its ability to store and share information – is the key to explaining and understanding many of today's so-called esoteric mysteries, especially 'past-life' experiences. As the Universe is replete with water, and the Earth much more so, the 'memory' of water also explains how the Astral Reality (Veil/Shadow) has been created over time by our imbalanced thoughts and beliefs, our fears, and the need of our 'leaders' to control people through ego and greed. It has been due to these imbalanced thoughts and desires of our lawmakers over time and the beliefs they have imposed upon us to control us that we have collectively (as a species) created this imbalanced external reality, and as time has progressed, this astral reality has grown even more powerful; to such an extent, that it is now much more influential in controlling our belief systems today.

Water stores and shares information from, and with, ALL living entities. Therefore, all our beliefs, fears, hopes, imaginations and thoughts have been stored in the Earth's consciousness and beyond, and these 'memories' are then shared (reflected) back to us on a soul/consciousness (subliminal) level through our natural interaction with everything around us. Living entities with a high-water content, such as human beings, are more easily influenced by this external reality, and as a consequence, readily buy into its deception becoming 'Channellers' of so-called spiritual wisdom.

Water reflects our past mistakes back to us so we can learn from them, not so we can take them on board as some form of so-called spiritual truth, make money, feed our egos, and in so doing, *continue* to spread this disease in consciousness. We've just forgotten that's all.

18

Water Memory[1]

Craig Hoffman

Editor's Introduction: Research by the author indicates that water carries the memory of its contact with the surroundings in the form of molecular bonds.

How is it that water can retain the memory of something that has been dissolved in it, even if that something has been filtered out? What is this miraculous property? It is called hydrogen-bonding, and it is nothing short of a miracle. According to science, water should be a gas-not a liquid. Yet, this simple hydrogen-bond, the loose connection between adjoining water molecules, prevents it from becoming a gas. Known science says 'water should not do this' as it does not occur in similar fluids. Known science has also proven that bumble-bees cannot fly…Of course, this merely means Mother Nature knows more than current science—she always will.

This same property that prevents water from being a gas, also gives water a form of memory. Because there is also another form of water memory, I term this particular type, Macro-Memory. This is because it depends on many interconnected water molecules working in unison, behaving more like a macromolecule. To understand Macro-Memory, we

1. Adapted from: Water Memory, Part One: Macro-Memory, Monday, February 8, 2010, http://incleanhealth.com/blog/water-memory-macro-memory/

first need to understand what causes hydrogen-bonding and also what happens when something is dissolved in water.

Hydrogen-bonding is the result of water being a dipole molecule-that is, it has two poles. This means that on the physical molecule, some areas have a positive electrical charge while other areas have a negative electrical charge. Just as a pile of small magnets (each with a north and south pole) would all connect together into a big clump, water molecules connect to one another. Specifically it is the oxygen atom of the water molecule that maintains the negative charge and two hydrogen atoms that maintain the positive charge. An oxygen atom is also much, much larger than a hydrogen atom. In this way, the positively charged hydrogen atoms of water molecules are attracted to and become loosely attached to the large oxygen atom, with lots of connecting space, of other water molecules (Fig. 1).

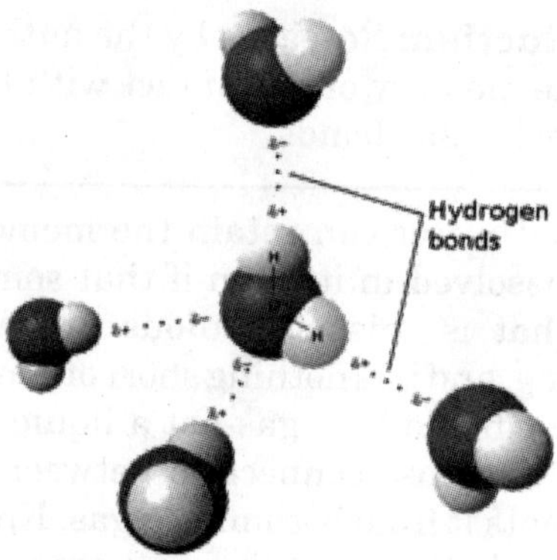

Fig. 1.

We think of water as simply being H_2O, but really water is only water when two or more molecules are bound together in hydrogen-bonds.

Now that we understand hydrogen-bonding is a result of water being a dipole molecule, we can also understand what happens when something is dissolved in water. Let us use common table salt as an example.

Common table salt has the chemical formula NaCl, sodium chloride. It is one sodium atom bound to one chlorine atom. Most of the time salt is a solid, forming a lovely crystalline structure.

It is beyond the scope of this article to discuss ionic bonds, the phenomenon that allows the sodium and chlorine to join together. Let us just say for this discussion that sodium, having no electrical charge, gives up an electron to the chlorine who wants it more. Since sodium lost an electron (which has a negative electrical charge), the sodium now has a positive charge and is chemically written as Na^+. Since the chlorine molecule now has the extra electron, it now has a negative charge, Cl^-. We know what happens when something has a positive charge and something has a negative charge, do we not? They attract, and this particular bond as a result of this attraction is called an ionic bond. This ionic bond was explained in a series of steps, but truthfully the sodium is never without the electron as it shares it with the chlorine. However, the above explanation will help you better understand what happens when salt is dissolved in water.

Water begins to penetrate the crystalline structure of the bulk salt, separating the individual molecules. Then it begins to work its way in between the sodium and the chlorine that make up the individual molecules. As mentioned above, chlorine has a high affinity for the electron, so as the water works its way between the sodium and the chlorine, the chlorine keeps the shared electron (Fig. 2).

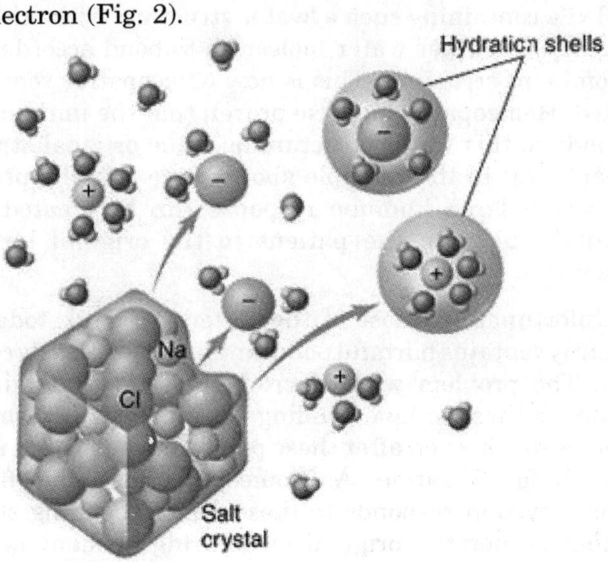

Fig. 2.

Now we no longer have salt, but two separate ions, Na^+ and Cl^-. One would tend to think that the Na^+ and Cl^- would immediately want to come back together because of their opposite charges, but they cannot sense each others' charge because water is an insulator. Being an insulator means it does not allow for the ready transfer or sharing of electrons. This insulating layer of water that immediately surrounds the newly formed ions, is termed hydration.

Now, let us look at how Macro-Memory works. Water as a dipole molecule, will not surround that which is dissolved in it randomly. For instance, Na^+ has a positive charge, so it is the negatively charged oxygen atom of the water molecule that moves in close. Na^+ has its own unique size and charge-density, so only a finite number of water molecules (parked oxygen first) can surround it. This first layer of water molecules to bond with a dissolved substance is called the 'hydration layer'. Then, according to the dipole effect of hydrogen-bonding, other water molecules bond in turn with the water molecules that formed the hydration layer.

In homeopathy, it has been found that simply shaking a small vile containing such a 'water structure' with other water, will cause the other water molecules to bond according to the predominant structure. This is how homeopathic remedies are created. Homeopathy has also proven that the immune system responds to this water structure as if the original structuring element (Na^+ in the example above) were actually present. In this way a large immune response can be created without actually subjecting the patient to the original structuring element.

Unfortunately, most of the water we drink today either presently contains harmful pollutants or very recently contained them. The problem with Macro-Memory is that fairly large portions of these unique bonding structures are present in the water we drink, even after these pollutants have been removed-so much for filtration. As homeopathy has shown us, the immune system responds to these unique bonding structures whether or not the original structuring element is actually present.

It seems the longer water has bonded in a particular structure, the more apt it is to re-bond in that same way, even if momentarily altered (a Micro-Memory effect). Such is the case with standard purification. Distillation removes most Macro-Memory, but the re-condensed water immediately picks up a new structure based on the construction materials within the machine, mostly metals and plastics. Have you ever tasted distilled water? It generally has a metallic and plastic taste. A similar scenario emerges from Reverse Osmosis devices.

The most reliable method for erasing Macro-Memory, is also the method used by Mother Nature—the spiraling vortex action in fast-flowing rivers and streams.

Fig. 3.

Coincidently, it was recently discovered that blood moves in a spiraling action in veins and arteries.

This vortex action simultaneously stretches the existing structure while adding energy. This added energy locks in the new structure, permanently erasing Macro-Memory.

Conclusion is that water is a grand store of memory.

19

Living Water[1]

Alex Andersson

Editor's Introduction: Viktor Schauberger researched natural flowing rivers and concluded that the swirling motion of free-flowing rivers gave it special energy. Water that is stationary or which is flowing through pipes loses this energy. This indicates that hydropower projects lead to loss of energy of water.

Viktor Schauberger's central theme was: 'Prevailing technology uses the wrong forms of motion. Our machines and processes channel such agents as air, water and other liquids and gases into the type of motion which Nature only uses to decompose and dissolve matter. Nature uses another form of motion for rebuilding. When our technology only uses the decomposing motion, it becomes a dead technology, a destructive one, dangerously affecting all of Nature.'

Viktor Schauberger wished to make practical use of nature's reconstituting principle of 'cycloid spiral motion'. In this he succeeded in some ways, though not in others. His description of what he experienced in Nature can sometimes sound like pure fantasy. However, it should be remembered that much of what Viktor Schauberger studied for so many years cannot now

1. http://www.archive.org/stream/AlexAndersson-LivingWater/Alexandersson-Living Water-Viktor Schauberger and the Secrets of Natural Energy 1990_djvu.txt.

be seen, for the natural environment in which he lived is now no more. Nothing alters the fact that the phenomena he describes could have naturally occurred in the unspoiled surroundings of his younger days, and that they have now disappeared through the environmental destruction that has broken down Nature's inter-connectedness.

> In Hetzau, below Ring, lie the Od Lakes (Odseen). After a long spell of hot weather a thunder-like noise (buhlen in the local dialect) is often heard coming from the bottom of the lake, accompanied by a water spout. One hot summer day I sat on the bank of the lake and wondered whether I should cool down by taking a refreshing bathe. Just as I decided to jump in, I noticed the water beginning to move in peculiar spiral whorls. Trees, which had been dumped in the lake by avalanches, began to describe a sort of spiral dance, which drew them constantly nearer, with ever increasing speed, to the centre of the lake. Having reached the middle, the trees suddenly took up a vertical position and then appeared to be sucked down into the depths by some dragging force, causing the bark to be ripped off. It could be likened to the experience of a man suddenly hurled upwards in the air by a cyclone, to crash down to earth stark naked. No tree reappeared from out of the Od lake.
>
> In a short time, the lake was again calmer, as if it had been freed by the victims which had been dragged down into the depths. It was, however, only the calm before the real storm. Suddenly the bed of the lake began to rumble. Without warning, a water spout of at least the height of a house shot upwards from the middle of the lake. A noise like thunder accompanied the turning cuplike pillar of water. Then, as suddenly, the spout collapsed upon itself. Waves hit the banks of the lake as the water began to rise in a mysterious way, and I was forced to leave hurriedly. I had experienced the archetypal expansion of water, a renewal of water in the lake, without any inflow.

Schauberger draws an audacious conclusion from his experience by the lake. According to his theories on water, it is a living substance which is born and develops—normally to change into higher forms of energy-but can, with incorrect treatment, also die. Even a restricted volume of water can increase, not in the usual sense of expansion through heat, but instead through growth like an organism. Schauberger continues:

> Naturally moving water augments itself. It improves its quality and matures considerably. Its boiling and freezing points change, and wise Nature makes use of this phenomenon to raise water, without using pumping equipment, to the highest mountain peaks, to appear as mountain springs. This conception of raising water is not to be taken literally, since in this context it is concerned with the natural process of propagation and purification. This in turn helps towards the expansion of air by creating an air cover, which serves to develop a higher form of life.

Schauberger now began to attempt to reproduce these stages. He built the first so-called water refining apparatus around 1930, and finally developed a model for which he sought a patent.

He started with sterilized water from the Danube, added small measures of certain metals, minerals and carbon dioxide. Sterilised water from container in mixed drop by drop with the salt solution. The mixture is then sprayed out from a perforated pipe, while carbon dioxide is being introduced. Water falls in droplets to the bottom while absorbing carbon dioxide, and is forced into a meandering motion where it passes over gold and silver filaments, to finally gather and cool in the silver lined container, until it slowly reaches +4°C. The whole process was an attempt to copy water's natural journey in the earth as he understood of its 'full cycle'. After a short storage period the water was allowed slowly to increase in temperature to 8°C, and was then ready to drink.

Rumour soon spread that Viktor Schauberger could make 'living water' and people streamed to his home to try it. The general opinion was that the water was very refreshing; the sick felt better, fevers abated and recovery quickened. Schauberger had already been nicknamed 'water magician' and now he was really thought to be one. Specimens sent to laboratories for analysis showed that Schauberger's water could not be differentiated from spawater.

Schauberger's proposals for the natural treatment of water were most controversial:

(1) Water must be allowed to flow and mature in its own natural environment, which, amongst other things, presupposes a naturally-grown forest containing a great variety of species. Both single crop forestry and clear felling must cease.

(2) All watercourses, from the little stream to the mature river, must have banks grown with trees and bushes to give natural shade.

(3) Water installations (dams, power stations, etc.) must be sympathetic to water's needs and must not alter its natural forms of motion.

(4) Water pipes, and other water transporting methods, must be so designed and of such material as to promote the preservation and development of water's particular biological quality.

20

Water Memory: Implications from Homeopathy[1]

Wikipedia

Editor's Introduction: French immunologist Jacques Benveniste proposed that water molecules rearrange themselves once a material is added to water. The new molecular arrangement persists even after the material responsible is removed. This indicates that water has an inherent ability to 'remember' the past influences. This belief underlies much of homeopathy.

This article taken from Wikipedia details the debate. It shows that there is no agreement among scientists on this issue. However, the very fact that this debate persists indicates that perhaps we do not have the scientific apparatus to measure and recognize these impacts. In the next chapter a confirmation of Benveniste in reported.

Water memory is a speculation that water is capable of retaining a "memory" of substances once dissolved in it.[1, 2] Shaking the water at each stage of a serial dilution is claimed to be necessary for an effect to occur.[3] The concept was proposed by Jacques Benveniste to explain the purported therapeutic

1. http://en.wikipedia.org/wiki/Water_memory.

powers of homeopathic remedies, which are prepared by diluting solutions to such a high degree that not even a single molecule of the original substance remains in each final preparation. Benveniste sought to prove this basic tenet of homeopathy by conducting an experiment to be published "independently of homeopathic interests" in a major journal.[4]

While some studies, including Benveniste's, have reported such an effect, double-blind replications of the experiments involved have failed to reproduce the results, and the concept is not accepted by the scientific community.[5] Liquid water does not maintain ordered networks of molecules longer than a small fraction of a nanosecond.[6]

The most prominent advocate of this idea was the French immunologist Jacques Benveniste. His team at the French National Institute of Health and Medical Research (INSERM) diluted a solution of human antibodies to such a degree that there was virtually no possibility that a single molecule remained. Nonetheless, they reported, human basophils responded to the solutions just as though they had encountered the original antibody (part of the allergic reaction). The effect was reported only when the solution was shaken violently during dilution. Benveniste stated: "It's like agitating a car key in the river, going miles downstream, extracting a few drops of water, and then starting one's car with the water." [7] At the time, Benveniste offered no theoretical explanation for the effect.

Benveniste submitted the research to the prominent science journal *Nature* for publication. There was concern on the part of *Nature's* editorial oversight board that the material, if published, would lend credibility to homeopathic practitioners even if the effects were not replicable. There was equal concern that the research was simply wrong, given the changes that it would demand of the known laws of physics and chemistry. The editor of *Nature*, John Maddox, stated that, "Our minds were not so much closed as unready to change our whole view of how science is constructed." Rejecting the paper on any objective grounds was deemed unsupportable, as there were no methodological flaws apparent at the time.

In the end, a compromise was reached. The paper was published in *Nature*, Vol. 333 on 30 June 1988, but it was accompanied with an editorial by Maddox that noted "There are good and particular reasons why prudent people should, for the time being, suspend judgment" and described some of the fundamental laws of chemistry and physics which it would violate, if shown to be true. Additionally, Maddox demanded that the experiments be re-run under the supervision of a hand-picked group of what became known as "ghostbusters", including Maddox, famed magician-cum-paranormal researcher James Randi, and Walter Stewart, a physicist and free-lance debunker at the US National Institutes of Health.

In the first series of supervised experiments, the original experimental procedure was followed as it had been when the paper was first submitted for publication. The experiments were successful, matching the published data quite closely. However, Maddox noted that during the procedure the experimenters were aware of which test tubes originally contained the antibodies and which did not. A second experimental series was started with Maddox and his team in charge of the double-blinding; notebooks were photographed, the lab videotaped, and vials juggled and secretly coded. Randi went so far as to wrap the labels in tinfoil, seal them in an envelope, and then stick them on the ceiling so Benveniste and his colleagues could not read them. No memory effect was observed in the blinded experiments.

Nature published a follow-up report in the next issue:[8] "We conclude that there is no substantial basis for the claim that antibody at high dilution (by factors as great as 10^{120}) retains its biological effectiveness, and that the hypothesis that water can be imprinted with the memory of past solutes is as unnecessary as it is fanciful." Nevertheless, there was no suggestion of fraud; Maddox and his team initially speculated that someone in the lab "was playing a trick on Benveniste," but later concluded, "We believe the laboratory has fostered and then cherished a delusion about the interpretation of its data." Maddox also pointed out that two of Benveniste's researchers were being paid for by the French homeopathic company Boiron.

In a response letter published in the same issue of the journal, Benveniste lashed out at Maddox and complained about the "ordeal" he endured at the hands of the *Nature* team, comparing it to "Salem witchhunts or McCarthy-like prosecutions."[9] In both the *Nature* response and a following *Quirks and Quarks* episode, Benveniste especially complained about Stewart, who he stated acted as if they were all frauds and treated them with disdain, complaining about his "typical know-it-all attitude". In his *Nature* letter, Benveniste also implied that Randi was attempting to hoodwink the experimental run by doing magic tricks, "distracting the technician in charge of its supervision!" He was more apologetic on *Quirks and Quarks*, re-phrasing his mention of Randi to imply that he had kept the team *amused* with his tricks and that his presence was generally welcomed. He also pointed out that although it was true two of his team-members were being paid for by a homeopathic company, the same company had paid for Maddox's team's hotel bill.

Maddox was unapologetic, stating "I'm sorry we didn't find something more interesting." On the same *Quirks and Quarks* show he dismissed Benveniste's complaints, stating that the possibility that the results would be unduly promoted by the homeopathy community demanded an immediate re-test. In failing, the tests demonstrated that the initial results were likely due to the experimenter effect. He also pointed out that the entire test procedure that Benveniste later complained about was one that had been agreed upon in advance by all parties. It was only when the test then failed that Benveniste disputed its appropriateness.

The debate continued in the letters section of *Nature* for several issues before being ended by the editorial board. It continued in the French press for some time.[10] For all of the arguing over the retests, it has done nothing to stop what Maddox worried about; even in the light of their failure they are still used to claim that the experiments "prove" that homeopathy works.[11] One of Benveniste's co-authors on the *Nature* paper, Francis Beauvais, later stated that while unblinded experimental trials usually yielded "correct" results (*i.e.* ultradiluted samples were biologically active, controls were not), "the results of blinded samples were almost always at random and did not fit the expected results: some 'controls' were active and some 'active' samples were without effect on the biological system."[12]

After the *Nature* controversy, Benveniste gained the public support of Brian Josephson,[13] a Nobel laureate physicist with a reputation for openness to paranormal claims. Experiments continued along the same basic lines, culminating with a 1997 paper claiming the effect could be transmitted over phone lines.[14] This was followed by two additional papers in 1999[15] and another on remote-transmission in 2000 by which time it was claimed that it could also be sent over the internet.[16]

The Time magazine reported in 1999 that, in response to skepticism from physicist Robert Park, Josephson had challenged the American Physical Society (APS) to oversee a replication by Benveniste. This challenge was to be "a randomized double-blind test", of his claimed ability to transfer the characteristics of homeopathically diluted water over the Internet.[17] The APS accepted the challenge and offered to cover the costs of the test. When he heard of this, Randi also offered to throw in the long-standing $1 million prize for any positive demonstration of the paranormal, to which Benveniste replied: "Fine to us."[18] in his *DigiBio News Letter*. However, Randi later noted that Benveniste and Josephson did not follow up on their challenge, mocking their silence on the topic as if they were missing persons.[19]

An independent test of the 2000 remote-transmission experiment was carried out in the USA by a team funded by the United States Department of Defense. Using the same experimental devices and setup as the Benveniste team, they failed to find any effect when running the experiment. Several "positive" results were noted, however, but only when a particular one of Benveniste's researchers was running the equipment. "We did not observe systematic influences such as pipetting differences, contamination, or violations in blinding or randomisation that would explain these effects from the Benveniste investigator. However, our observations do not exclude these possibilities."

Benveniste admitted to having noticed this himself. He stated that "certain individuals consistently get digital effects and other individuals get no effects or block those effects." The experiment is notable for the way it attempted to avoid the confrontational nature of the earlier Maddox test.[20]

Third-party attempts at replication of the Benveniste experiment have failed to produce positive results that could be independently replicated. In 1993, *Nature* published a paper describing a number of follow-up experiments that failed to find a similar effect,[21] and an independent study published in *Experientia* in 1992 showed no effect.[22] An international team led by Professor Madeleine Ennis of Queen's University of Belfast claimed in 1999 to have replicated the Benveniste results.[23, 24] Randi then forwarded the $1 million challenge to the BBC *Horizon* program to prove the "water memory" theory following Ennis' experimental procedure. In response, experiments were conducted with the Vice-President of the Royal Society, Professor John Enderby, overseeing the proceedings. The challenge ended with no memory effect observed by the Horizon team.[25] For a piece on homeopathy, the ABC program 20/20 also attempted, unsuccessfully, to reproduce Ennis's results.[26]

Notes

1. Anonymous [John Maddox] (1988). "When to believe the unbelievable". *Nature* 333 (6176): 787.
2. See Mole (unit) and Homeopathy for more detailed information on how we can calculate the original number of molecules.
3. Dayenas E., Beauvais F., Amara J., Oberbaum M., Robinzon B., Miadonna A., Tedeschit A., Pomeranz B., Fortner P., Belon P., Sainte-Laudy J., Poitevin B. and Benveniste J. (30 June 1988). "Human basophil degranulization triggered by very dilute antiserum against IgE" (PDF). *Nature* 333 (6176): 816–818. http://br.geocities.com/criticandokardec/benveniste01.pdf. Retrieved 2007-06-05.
4. Poitevin, Bernard (2005). "Jacques Benveniste: a personal tribute". *Homeopathy* 94 (2): 138–139.
5. P. Ball (8 August 2007). "Here lies one whose name is writ in water". *Nature*. doi:10.1038/news070806-6. http://

www.nature.com/news/2007/070806/full/news070806-6.html.

6. Cowan ML, Bruner BD, Huse N, *et al.* (2005). "Ultrafast memory loss and energy redistribution in the hydrogen bond network of liquid H_2O". *Nature* 434 (7030): 199–202. doi:10.1038/nature03383. PMID 15758995.
7. Langone John (8 August 1988). "The Water That Lost Its Memory". *Time Magazine*. http://www.time.com/time/magazine/article/0,9171,968080,00.html. Retrieved 2007-06-05.
8. Maddox J.; Randi J., Stewart W. W. (28 July 1988). "High-dilution experiments a delusion" (PDF). *Nature* 334: 287–290. doi:10.1038/334287a0. http://br.geocities.com/criticandokardec/benveniste02.pdf. Retrieved 2007-06-05.
9. Benveniste J. (28 July 1988). "Dr Jacques Benveniste replies" (PDF). *Nature* 334: 291. doi:10.1038/334291a0. http://br.geocities.com/criticandokardec/benveniste02.pdf. Retrieved 2007-06-05.
10. Coles P. (28 July 1988). "Benveniste controversy rages on in the French press". *Nature* 334 (6181): 372. doi:10.1038/334372a0. PMID 2457165.
11. Homeopathy breakthrough.
12. Memory of water and blinding, Francis Beauvais, *Homeopathy*, 97(1):41-42, January 2008.
13. Josephson Brian, molecule memories, *New Scientist* letters, 1 November 1997
14. Benveniste J.; Jurgens P., Hsueh W. and Aissa J. (21–26 February 1997). "Transatlantic Transfer of Digitized Antigen Signal by Telephone Link". *Journal of Allergy and Clinical Immunology*.
15. Benveniste J.; Aissa, J., Guillonnet. "The molecular signal is not functional in the absence of "informed water"". *Medical Hypotheses* 54 (A163 (abstr.)).
16. Benveniste J.; Thomas Y, Schiff M, Belkadi L, Jurgens P, Kahhak L (2000). "Activation of human neutrophils by electronically transmitted phorbol-myristate acetate". *FASEB Journal* 13 (1): 33–39. doi:10.1054/mehy.1999.0891. PMID 10790721.
17. Leon Jaroff, Homeopathic E-Mail, *Time Magazine*, 17 May 1999].

18. Jacques Benveniste and Didier Guillonnet, DigiBio - NewsLetter 1999.2, "Demonstration challenge, etc." section.
19. James Randi, Computer problems, a Nobel laureate reneges, more magnetic shoes, the metric system, and..., *Commentary*, 26 January 2001.
20. Jonas, Wayne B.; John A. Ives, Florence Rollwa-gen, Daniel W. Denman, Kenneth Hintz, Mitchell Hammer, Cindy Crawford, and Kurt Henry (January 2006). "Can specific biological signals be digitized?". *FASEB Journal* 20 (1): 23–28.doi:10.1096/fj.05-3815hyp. PMID 16394263. http://www.fasebj.org/cgi/content/full/20/1/23. Retrieved 2007-06-05. —this paper includes an excellent references list.
21. Hirst S. J.; Hayes N. A., Burridge J., Pearce FL, Foreman JC. (9 December 1993). "Human basophil degranulation is not triggered by very dilute antiserum against human IgE". *Nature* 366 (5): 525–527. doi:10.1038/366525a0. PMID 2455231.
22. Ovelgonne, J. H.; Bol, A. W., Hop, W. C., van Wijk, R (15 May 1992). "Mechanical agitation of very dilute antiserum against IgE has no effect on basophil staining properties". *Experientia* (Birkhäuser Verlag (part of Boiron) 48 (5): 504–508. doi:10.1007/BF01928175. PMID 1376282.
23. Belon P.; Cumps J., Ennis M., Mannaioni P. F., Sainte-Laudy J., Roberfroid M., Wiegant F. A. C. (April 1999). "Inhibition of human basophil degranulation by successive histamine dilutions: Results of a European multi-centre trial". *Inflammation Research* 48 (Supplement 1): 17–18. doi:10.1007/s000110050376.
24. Lionel Milgrom (2001-03-15), "Thanks for the memory. Experiments have backed what was once a scientific 'heresy', says Lionel Milgrom", *The Guardian*, http:// www. guardian.co.uk/science/2001/mar/15/technology2
25. "Homeopathy: The test. Transcript.". BBC Two. 2003-11-26, 9 pm. http://www.bbc.co.uk/science/horizon/ 2002homeopathytrans. shtml. Retrieved 2007-03-04.
26. Stossel, John (2008). "*Homeopathic Remedies - Can Water Really Remember?*". 20/20 *(ABC News). http://abcnews.go.com/print?id=124309. Retrieved 2008-01-22.*Retrieved from "http://en.wikipedia.org/wiki/ Water_ memory".

21

Korean Scientists Confirm Water Memory[1]

Andy Coghlan

Editor's Introduction: In the previous Chapter the debate revolving around the theories of Jacques Benveniste has been outlined. In this chapter, taken from *New Scientist*, Benveniste's theories appear to find support from a team of scientists from Korea.

A team in South Korea has discovered a whole new dimension to just about the simplest chemical reaction in the book—what happens when you dissolve a substance in water and then add more water.

Conventional wisdom says that the dissolved molecules simply spread further and further apart as a solution is diluted. But two chemists have found that some do the opposite: they clump together, first as clusters of molecules, then as bigger aggregates of those clusters. Far from drifting apart from their neighbours, they got closer together.

The discovery has stunned chemists, and could provide the first scientific insight into how some homeopathic remedies

1. Adapted from: "Bizarre chemical discovery gives homeopathic hint," New Scientist, November 2001.

work. Homeopaths repeatedly dilute medications, believing that the higher the dilution, the more potent the remedy becomes.

Some dilute to "infinity" until no molecules of the remedy remain. They believe that water holds a memory, or "imprint" of the active ingredient which is more potent than the ingredient itself. But others use less dilute solutions - often diluting a remedy six-fold. The Korean findings might at last go some way to reconciling the potency of these less dilute solutions with orthodox science.

Completely Counterintuitive

German chemist Kurt Geckeler and his colleague Shashadhar Samal stumbled on the effect while investigating fullerenes at their lab in the Kwangju Institute of Science and Technology in South Korea. They found that the football-shaped buckyball molecules kept forming untidy aggregates in solution, and Geckler asked Samal to look for ways to control how these clumps formed.

What he discovered was a phenomenon new to chemistry. "When he diluted the solution, the size of the fullerene particles increased," says Geckeler. "It was completely counterintuitive," he says.

Further work showed it was no fluke. To make the otherwise insoluble buckyball dissolve in water, the chemists had mixed it with a circular sugar-like molecule called a cyclodextrin. When they did the same experiments with just cyclodextrin molecules, they found they behaved the same way. So did the organic molecule sodium guanosine monophosphate, DNA and plain old sodium chloride.

Dilution typically made the molecules cluster into aggregates five to 10 times as big as those in the original solutions. The growth was not linear, and it depended on the concentration of the original.

"The history of the solution is important. The more dilute it starts, the larger the aggregates," says Geckeler. Also, it only worked in polar solvents like water, in which one end of the molecule has a pronounced positive charge while the other end is negative.

Biologically Active

But the finding may provide a mechanism for how some homeopathic medicines work—something that has defied scientific explanation till now. Diluting a remedy may increase the size of the particles to the point when they become biologically active.

It also echoes the controversial claims of French immunologist Jacques Benveniste. In 1988, Benveniste claimed in a *Nature* paper that a solution that had once contained antibodies still activated human white blood cells. Benveniste claimed the solution still worked because it contained ghostly "imprints" in the water structure where the antibodies had been.

Other researchers failed to reproduce Benveniste's experiments, but homeopaths still believe he may have been onto something. Benveniste himself does not think the new findings explain his results because the solutions were not dilute enough. "This [phenomenon] cannot apply to high dilution," he says.

Fred Pearce of University College London, who tried to repeat Benveniste's experiments, agrees. But it could offer some clues as to why other less dilute homeopathic remedies work, he says. Large clusters and aggregates might interact more easily with biological tissue.

Double-check

Chemist Jan Enberts of the University of Groningen in the Netherlands is more cautious. "It's still a totally open question," he says. "To say the phenomenon has biological significance is pure speculation." But he has no doubt Samal and Geckeler have discovered something new. "It's surprising and worrying," he says.

The two chemists were at pains to double-check their astonishing results. Initially they had used the scattering of a laser to reveal the size and distribution of the dissolved particles.

To check, they used a scanning electron microscope to photograph films of the solutions spread over slides. This, too, showed that dissolved substances cluster together as dilution increased.

"It doesn't prove homeopathy, but it's congruent with what we think and is very encouraging," says Peter Fisher, director of medical research at the Royal London Homeopathic Hospital.

"The whole idea of high-dilution homeopathy hangs on the idea that water has properties which are not understood," he says. "The fact that the new effect happens with a variety of substances suggests it's the solvent that's responsible. It's in line with what many homeopaths say, that you can only make homeopathic medicines in polar solvents."

Geckeler and Samal are now anxious that other researchers follow up their work. "We want people to repeat it," says Geckeler. "If it's confirmed it will be groundbreaking".

22

Water: Essential for Existence[1]

Dan Stewart and Denise Routledge

Editor's Introduction: This article explains that water in free flow has more energy which is lost when water flows through pipes. Implication is that river water loses energy when made to flow through tunnels for generation of hydropower.

Hans Kronberger, author of "On The Track of Water's Secret," states: "Injuries to nature begin in small ways; for example, tap water that is supplied to consumers via conventional pipe lines. According to naturalists, water loses its original energy through friction in the pipes and straight line of transport... it was not a coincidence that Romans built long, open air water transport systems with winding construction, using natural materials: wood and natural stone."

"In normal conditions," Austrian inventor and naturalist Jahann Grander maintains, "Water flows either on the surface of the earth or deep under-ground, always seeking a natural course. In our water supply systems, however, the water is collected and forced through pipelines under pressure. At this stage the water suffers a serious aggression for the first time. The high pressures...are highly detrimental to the liquid. The water is then further contaminated by the addition of powerful

1. Adapted from Explore, Volume 8, Number 5, 1998.

chemicals such as chlorine. However we use it, the water eventually finds its way back to nature through the drains. We collect clean water from nature and return it soiled and sick."

"Most of the people today have lost every bit of knowledge about what really happens in nature," states Grander. "That is why I wish people themselves would start to think about the processes of nature again and through this learn to respect nature again." Grander feels this knowledge has been lost "because in times of constant economic growth and constant technical progress people try to convince us that scientific research offers solutions that make observation of the laws of nature superfluous."

After decades of research Grander has succeeded in imprinting natural and Universal energies back into our water. Implosion is a key point in the science of revitalizing of water, or the erasing of information. But what is implosion? Quite simply, it is the opposite of explosion. "All of our contemporary technologies and the supply of energy are based on explosion, or 'expansion' by generating heat by burning raw or refined materials," states Hans Kronberger, author, On The Track of Water's Secret: "Explosion and expansion are centrifugal forces, while implosion works inward, or centripetally. Implosion concentrates its force into the center, where it then becomes the strongest." Nature carries this system out to perfection. We can observe these processes in motions that take place in nature.

How can the high positive energy of nature's water be maintained when water is confined to restrictive systems? Grander has studied the effects nature has on water for many years. "Water needs freedom," he states. "Living water seeks out its source of energy itself. It flows over the earth, inside the earth and makes long loops."

He claims that, "Not until we have recognised the perfect interplay and the mutual dependence of the four elements—earth, water, air and fire... can we imagine how destructive our interference with nature is and how much it disturbs the balance and order of things."

Water is the most essential element of life. It provides all life, whether plant, animal or human, with the essential energies of life. But where does water get these energies from?

The properties of water are perpetually changing. Naturalists have found that water transports energy and information of immense proportions. In nature, bodies of water that are left undisturbed by man turn within themselves, wind, bend and bubble, froth and foam, spiral and cascade. These untouched waters have the striking ability to rejuvenate waste that is added to it, and the water is revitalized. Water absorbs energy. Water carries energy. Water emits energy.

Johann Grander's work has evolved around three main principles: (i) water is a carrier of information, (ii) water has memory and (iii) negative information can be erased.

Johann Grander has developed a process which is able to renew the structure of water. His research has led him to encase an electromagnetically enhanced concentrate, which renews natures' own high energy vibrations to the structure of the water.

The Grander process changes the polarity from negative to positive during the process of dynamic implosion. Negative polarity is changed to positive and the amount of dissolved oxygen is increased. This allows harmful compounds to pass harmlessly through our bodies. The ordinary tap water transforms to resemble water found in a true mountain spring.

Dr Horst Felsch examined the Grander water for contents such as bacteria and nitrates to guarantee the water met the stringent standards set by Tyrolean laws. Originally, Dr Felsch could not make any sense of 'revitalised water' so he concentrated on the hygiene. He very quickly became curious. The logical question for the natural scientist was "does Grander water have a different energetic structure than normal drinking water, and if so, can it be proven. I think I was the first one to take a serious look at the microbiology of Grander water. I carried out routine germ count tests to check the quality."

Dr Horst Felsch states of his research into Grander's revitalized water: "it is known that liquids can be primarily activated in the way that part of the inner energy is transformed by means of refrigeration, so that a higher energy content is impressed. Like this, water in its electromagnetic structure has been successfully transformed in the way that—both by modifying the magnetic nuclear resonance properties (modification of the spin-spin coupling constants), and by the induction and formation of the super-molecular complexes between the single water molecules—the electromagnetic state of oscillation was modified. These electromagnetic oscillations are responsible for the formation of an electromagnetic field of defined frequency, phase and amplitude, which spreads evenly in all directions. Surprisingly enough, it turned out that, in such an electromagnetic field, larger particles and molecular groups are decomposed into smaller molecular groups, which results in a reduction of viscosity, the dissolution of harmful molecules and the like. And it is also highly surprising that the energetic state of the activated liquid is preserved, i.e., it does not ease off."

Felsch's study also included a germ count using the membrane filter method. In any normal water sample, the microorganisms (bacteria) are distributed unsymmetrically and chaotically on the membrane filter. In Grander's revitalised water, the microorganisms depart from their chance distribution and line up symmetrically.

Grander says that the process of re-energizing water is simply following the laws of nature. He says: "The fact that to so many people it seems to be a secret indicates how far science has already digressed from nature." Science, however, completely overlooks the fact that water—as a life carrier — is itself alive and needs to be kept in this condition if it is to fulfill its naturally ordained function.

The origin of life is water. *Water is life*. Immediate attention is required to rectify the problems we have created through carelessness, pollution and modernisation techniques. Revitalizing water is the solution to these concerns. Through revitalisation our water returns to the state that nature intended: clean, clear and complete with the vital energies created by Mother Nature. *Our very existence depends on it.*

Bibliography

1. Kronberger, Hans and Lattacher, Siegbert: On The Track of Water's Secret: Uranus, Vienna, 1995.

23

Molecular Resonance Effect Technology[1]

Howard Fisher

Editor's Introduction: Dr Fisher says that geomagnetic fields of earth affect the molecular structure of water. Natural water is more healthy because it is charged with these geomagnetic waves. He has developed technique to artificially treat water with waves similar to geomagnetic fields. He provides various proofs of the impact of such treatment upon properties of water.

Water of the Ganga flows below shrines of Badrinath. There are Holy Rocks here. The spiritual qualities of Ganga may arise from the geomagnetic field of these rocks.

Just recently the actual structure of naturally occurring water had been established. The formerly accepted theory of water as a "tetrahedrally (pyramidal) coordinated random network is now replaced by a structural organisation that instead strongly favours hydrogen-bonded water chains or large rings embedded in a weakly hydrogen-bonded disordered network."[1] Scientists now believe that this structure best explains some of the phenomenal abilities of water, but even so, the bonds between the molecules of water are very transitional,

1. Adapted from: Molecular Resonance Effect Technology: The Dynamic Effects on Human Physiology, Mothersill Press, Toronto, Canada, 2008.

lasting no more than a femto-second at best.[2] Water is constantly in transition. In other words, all the hype about clusters of five or six water molecules having particular properties now must rely upon some other explanation for the attributed properties.

Thanks to the work of the Japanese investigator Masaru Emoto,[3] we are beginning to understand that the structure of water can be changed, not only by the geomagnetic fields that vary diversely over the planet, but by the energy of thought as well. The bipolarity of water has been long established and the subsequent changes to bonding angles are not only plausible, but a regular occurrence. Using electromagnetic fields, there are subtle electrical effects that can be used to alter or modify the molecular structure of water and subsequently affect the properties and the ensuing physiological functions of this life-sustaining liquid that makes up more than seventy percent of the human body. We must consider the fact that since we know this does occur, some water will be more physiologically beneficial, and on the other hand, some will be less beneficial due totally to the changes in the water structure and the correlated changes in ensuing functions.

In 1986, Dr (Igor) Smirnov was part of the investigative team that was looking into the disaster at the Chernobyl nuclear power plant. There was a significant finding inside the affected perimeter: one group of irradiated people undergoing the rehabilitation in the health care facility in the area of mountain springs was not suffering from cancer, although millions of others were. This was the seed from which sprouted one of the most innovative molecular changes that we have seen to the structure of water.[4]

From his research into the Chernobyl disaster, Dr Smirnov determined that it was the specific molecular structure of the spring water from the Caucasus Mountains that helped people to counteract to the development of oncological diseases following the adverse effects from the radiation. This finding was not uncommon because weak electromagnetic fields have the ability to change the properties of water[5] and therefore geo-magnetic fields from different regions of the planet have the ability to affect the structure and subsequent physiological effects of water as well.

After researching the effects of electromagnetic radiation on cellular structures at St. Petersburg University in Russia, Dr Smirnov developed a system called Molecular Resonance Effect Technology (MRET) for the modification of the molecular organisation state of water and other liquid substances.

The theory behind MRET activated water is soundly vested in the laws of physics. The effect on electromagnetic force has on an atom depends on the atom's electronic structure. One could imagine that the application of the appropriate time-dependent force to an atom could alter its electronic structure in a specific way, thereby controlling its response to subsequent radiative or collisional processes. Furthermore, the specificity of certain reactions of the electronic structure might be used to reconstruct the motion of the atomic electron cloud. The key to the manipulation of electronic structure in atoms is the generation of electromagnetic fields or radiation that will push and pull the electronic wave function in a controlled and reproducible way. Thus, by controlling the dynamic parameters of the electromagnetic field it is possible to alter the structure of the fluid in a controlled manner, which may lead to the physiological effects of this fluid. The key is, of course, reproducibility.

It should come as no surprise to learn that what Dr Smirnov has done through his molecular resonance effect technology (MRET) has changed the actual molecular composition of water so that the structure and frequency are similar to intracellular water. Not only that, the electrical properties are modified and predispose a more ideal ion exchange (proton pump) across the cellular membrane. Add to this the decreased physiological viscosity and surface tension, and now we have a mechanism to deliver nutrients and bring out intracellular toxins. Nuclear magnetic resonance spectroscopy has shown this water to be structurally different post-activation as well as UV luminescence spectrophotometry.[6] The level of physiological activity of MRET water is high.

Furthermore, another of the premises behind molecular resonance effect technology is the direct transmission of pre-recorded molecular activity signals to biological systems (any person or animal) using the MRET activated water. One might consider these messages to be functional frequencies imprinted in water during the process of activation which may be carried out by the water with specifically modified long-range molecular structure. The subsequent effects of the activated water on bacteria, viruses, and abnormal cells, can be explained by the fundamental physical phenomenon of electromagnetism, such as resonance, constructive and destructive interference.[7]

How Does Activation Occur?

The water-activating device used in Molecular Resonance Effect Technology (MRET) is made of a polar polymer compound mixed with specific amounts of pharmacologically active organic and inorganic substances. During the activation process, the MRET polymer compound is placed within an externally generated distinct electromagnetic field. Specifically, the MRET polymer compound is exposed to a homogenous magnetic field and an oscillating optical light with a wavelength of 600-700 nanometres and a frequency of 7.8 Hertz.[8]

As a result, the MRET compound generates very subtle, low frequency electromagnetic oscillations similar to those healthy geomagnetic fields found in specific areas on Earth such as the Caucasus Mountains in Europe or the Tibetan Mountains in Asia where the waters have been shown to provide dynamic changes to physiological functions. The naturally occurring water in these regions has been shown for centuries to improve health, support healing, promote longevity, and help eliminate heavy metals, allergens and other toxins and from living organisms.

Consistent with the MRET theory, the applied electromagnetic field generates an excitation in the fractal geometry nano-ring structures of the MRET polymer compound. Due to the phenomenon of piezoelectricity and intensive electrical activity of the fractal nano-rings, this polymer generates biologically active subtle electromagnetic oscillations. The process of activation modifies the hydrogen-bonding patterns of water molecules and induces the formation of the long-range multilayer molecular structures compatible with the intracellular water structuring. These results were confirmed by experimental work described in this monograph.[9]

As complicated as the MRET activation process may seem, all you must realise is that the activation process does indeed restructure the water. The following findings will show the structural differences between the control water (source water) and the MRET activated water: dispersion staining microscopy, nuclear magnetic resonance spectrometry, high voltage photography and viscosity testing.

Dispersion Staining Microscopy

The use of a dispersion staining microscopy has shown differences in the images of the crystalline structures in frozen water samples of the control and the MRET activated water. The water is frozen with a liquid nitrogen injection and then examined under polarised light. The structure, thickness and incident angle of the light will affect the diffraction of light off the crystals.

The figures represented here are micrographs from testing conducted at Meixa Tech Laboratories by analyst Bryan Burnett. The crystalline structure of the frozen tap water shows extensive irregular fracturing and chaotic crystalline formation (Fig. 1). This may be due to the impurities and poorly organised molecular structure of the tap water which have interfered with the crystallisation process.

Fig. 1: Dispersion Staining Microscopy Tap Water.

The MRET activated water sample shows well organized crystalline formations with a strong tendency of the crystal axis to be oriented in one direction. This difference in the structure is a result of the modified molecular structure of MRET water (Fig. 2).

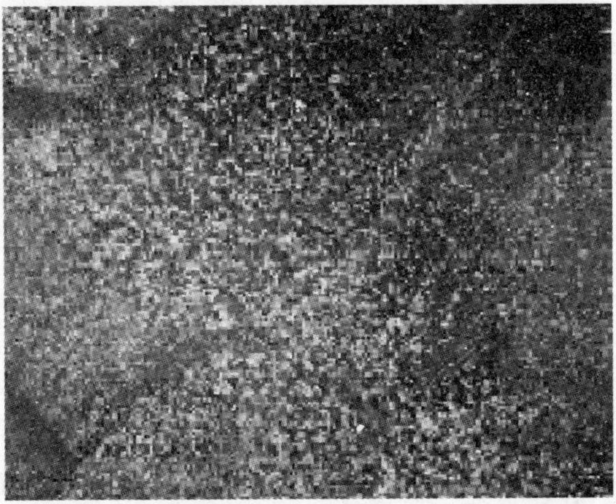

Fig. 2: Dispersion Staining Microscopy MRET Water.

Nuclear Magnetic Resonance

Nuclear magnetic resonance spectro-photometry has been shown to differentiate even the most subtle difference between molecules,[10] and therefore is an excellent modality to demonstrate the difference between MRET activated water and the control source water. Nuclear magnetic resonance imaging is an excellent tool for validating and determining differences in water structures.[11]

Dr Lin Chiang conducted the Nuclear Magnetic Resonance testing at NuMega Resonance Laboratory in San Diego, California. The test showed a consistent two and a half time (2.5-fold) increase in the width of the proton peak in the line of NMR absorption in MRET activated water (Figs. 3 and 4).[12]

The narrow proton peak of the control sample of water shown in Fig. 3 demonstrates a proton reactivity level that is extremely different from the results that are observed when testing the MRET activated water.

Regular Water

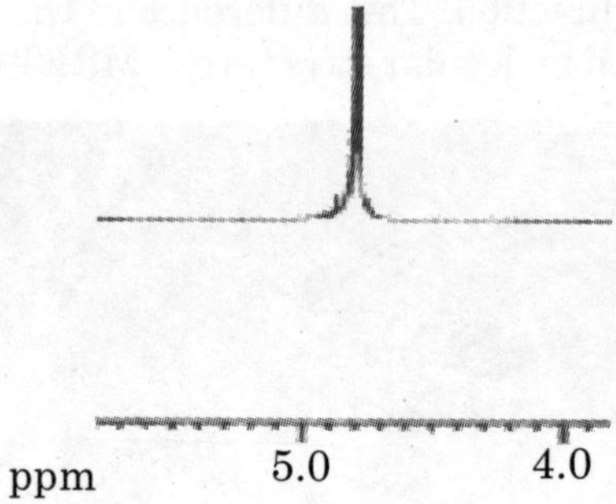

Fig. 3: Nuclear Magnetic Resonance Test Control.

The water sample shown in Fig. 4 was activated for thirty minutes. The increased width measured in the proton peak could only have been caused by the modification of the shape of molecules and molecular structure of the activated water. This demonstrates another tangible measurement where the difference between the initial control water and the activated MRET water can be visibly noted and reproduced upon subsequent NMR testing.[13] These results were produced by Dr. Leroy Lafferty in the Department of Chemistry laboratory at San Diego State University.[14]

Activated Water

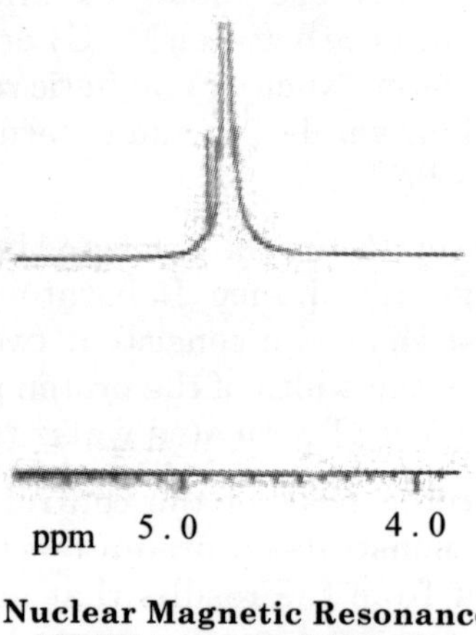

Fig. 4: Nuclear Magnetic Resonance Test MRET Activated Water.

High Voltage Photography

High voltage photography, sometimes referred to as Kirlian photography, is a method for detecting energy variations in an object. A high voltage is briefly applied to a discharge plate to make the photographic exposure. The corona discharge (electron flow) between the object and discharge plate passes through and is recorded on to the film. When the film is developed you have a high voltage photograph of the object. One of the prime uses of this photographic technique is to show changes in energy patterns.

The high-voltage photography showed that the MRET activated water demonstrated an enhanced Corona Discharge Effect – luminous fringes that appear around the electrically conductive samples of water (Figs. 5 and 6). The physical process of cold emission of electrons is responsible for producing the Corona Discharge phenomenon.

Regular Water

Fig. 5: High Voltage Photography—Regular Water.

Activated Water

Fig. 6: High Voltage Photography—Activated Water.

By examining the following two figures, the control water (before MRET activation) and the activated water, one can see that the increased emission of electrons in the activated water is visibly more intensive. The increase in the corona discharge effect is more than double, indicating that the water has a higher energy level.[15]

Since nothing was introduced to the water during the activation process to initiate the enhanced Corona increased electron release, it is reasonable to concede that proton activity in the MRET activated water has also increased.

Viscosity

Gathering information about water's viscosity and the subsequent physiological behaviour will allow us to understand the degree of difficulty or ease of the hydration or the permeation of the cellular membrane that will take place. Viscosity is defined as the measurement of the 'internal friction' of a fluid or the speed at which it will flow. The viscosity of any particular fluid, in this case water, defines the simplicity with which it may gain entry to the cell and to be used for absorption of nutrients or detoxification as well as for the hydration of the body. The relationship between hydration and aging has already been clearly established: aging is the progressive state of dehydration that occurs over time leading to an increase in cellular dysfunction. In other words, the more water we lose, the less efficiently we function.

The sophisticated scientific investigation showed that the viscosity of MRET activated water, when subjected to very low tangential pressure decreases up to 300-500 times compared to the non-activated water.[16] This research was conducted at Moscow State University, Russia using distilled water. It is scientifically proven that cellular processes in biological systems are driven by the low energy of the bio-chemical reactions inside and outside the cellular structures. Consequently, these processes create low tangential pressures along water surfaces and cellular membranes. Thus, the very low tangential pressures existing in biological systems contribute to the manifestation of the anomalously low viscosity of MRET water which results in improved intracellular/extracellular water exchange. This can contribute to the enhancement of the cellular transduction mechanism and the predisposition to optimal function of the cells in biological systems.[17]

The significance of this finding cannot be overstated. Water is the delivery system for nutrients and the major vehicle for removal of cellular wastes. By decreasing the viscosity of the water in the body structures, the potential for optimizing physiological activity has also been exponentially increased.

This decreased viscosity "means that the water has a 'super liquidity' characteristic and possesses very low resistance when it penetrates through small pores and capillaries in the body. Based on this fact, it is possible to conclude that MRET water may significantly improve the cellular membrane function as well as the physiological function of the circulatory system in the body."[18]

By using the verifiable, objective measurements of nuclear magnetic resonance spectrometry (increased width of the proton peak), viscosity testing under very low tangential pressure (300-500 times reduction in viscosity), high voltage photography (more than 100 per cent increase in corona discharge) and dispersion staining microscopy (visible structural change), there is a clearly demonstrable structural change that has occurred from the control (pre-activated) liquid to the resultant MRET activated liquid. The truly exciting phenomena about the effects of the activation process on water are the irrefutable dynamic kinetic changes, not only on human physiological responses to the water, but to the suspension or reversal of many potentially pathological processes.

MRET Activated Water Effects on Escherichia Coli

One of the most dangerous coliforms is Escherichia coli. The bacteria are very plentiful at the distal end of the digestive tract (colon) however it is extremely septic anywhere from the proximal end (mouth) to the stomach and can cause food poisoning and other infections. Fecal matter will contain E. coli.

A recent study on the effects of MRET activated water on Escherichia coli (E. coli) was conducted at the Kiyv (Kiev) Institute of Microbiology and Virology of Ukrainian Academy of Science in both aerobic and anaerobic environments. The aerobic study demonstrated a major reduction in the growth of these bacteria. By activating the growth medium for thirty minutes, E. coli growth was inhibited thirty times (30-fold). Nutrient medium activated for sixty minutes increased the effectiveness of the E. coli inhibition by a factor of ten (300-fold).[19]

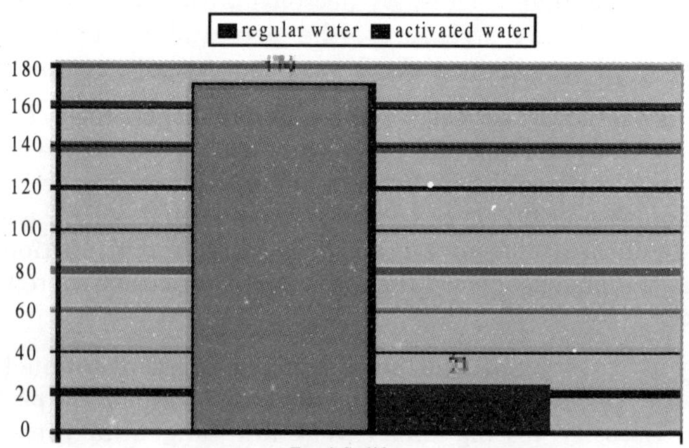

Fig. 7: MRET Activated Water Effects on Total Coliforms Count.

Acidic-alkaline Balance and MRET Water Memory

The pH scale is used to measure the acidity or alkalinity of any solution. It is a logarithmic scale that reflects the concentration of hydrogen ions in the solution expressed by reciprocals of the power of 10. The scale runs from zero to fourteen (0 to 14). Values less than seven (7) are in the acidic range; Values greater than seven (7) are in the alkaline range. A high pH value corresponds to a large negative power of 10, meaning a weaker concentration of hydrogen ions (alkaline). A low pH value indicates a small negative power of 10, which represents a high concentration (acid).

Pure water has a concentration of 10^{-7} hydrogen ions (H^+) which equals the concentration of hydroxyl groups (OH^-). This translates into a neutral pH factor of seven (7), which is neither acidic nor alkaline. Vinegar has a concentration of 10^{-4} hydrogen ions; its pH value is 4, which is acidic. Sodium hydroxide has a pH value of 14, which is alkaline extreme; hydrochloric acid has a pH value of 0, which is extremely acidic.

Investigative studies at Moscow State University, Russia and at CAI Environmental Laboratory in Carlsbad, California concerning pH revealed that the thirty minute MRET activation had a tendency to balance the pH values of water closer to neutral pH=7.0[20,21] compared to non-activated water. The results may depend on the initial pH value of the water, the type and the source of water, the time of activation and a number of environmental conditions.

The research, which was conducted at Moscow State University on distilled water which had been MRET activated for fifteen (15), thirty (30), forty-five (45) and sixty (60) minutes, as well as on non-activated water, revealed the oscillating and fluctuating character of pH values. The results for the first four hours after the water activation are presented on Fig. 8 and the results for the first sixteen days following the water activation are represented in Fig. 9.

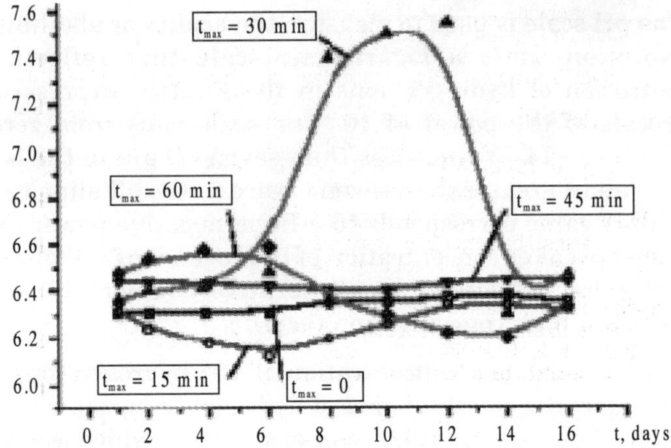

Fig. 8: MRET Water Effect on pH Values For The First Four Hours.

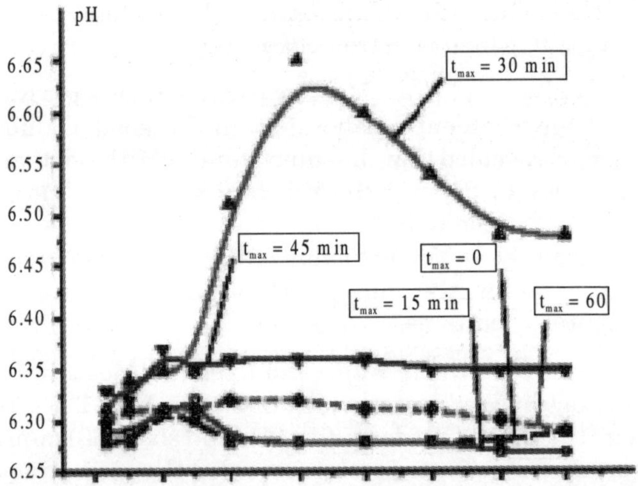

Fig. 9: MRET Water Effect on pH Values For The First 16 Days.

The fact that the oscillating changes in the pH value can be observed not only for the first hours but even during several days (at least fourteen days in case of thirty minutes of MRET activation) supports the idea that the water can keep MRET beneficial properties for several days at room temperature.

The results of this test confirm the scientifically valid phenomena of the existence of "water memory." In other words, the study verifies the ability of water to store information for a significant length of time. Another study involving oncological diseases in animal models conducted at Kiyv Institute of Experimental Pathology, Oncology and Radiobiology of Ukrainian Academy of Science confirmed the ability of MRET water to keep its health beneficial properties for forty-five (45) days when it was kept refrigerated at 0°C.

References

1. Head-Gordon T, Johnson M.E. Tetrahedral structure or chains for liquid water. *PNAS*. 2006; (103)21: pp. 7973-7977.
2. *Ibid*.
3. Emoto M, Thayne T A (translator). The Hidden Messages in Water. Beyond Words Publishing. Hillsboro, Ore 2004.
4. Smirnov I. Activated Water. *Electronic Journal of Biotechnology*. 2003; (6)2: pp.128-142.
5. Semikhina L P, Kiselev V F. Effect of weak magnetic fields on the properties of water and ice. *Russian Physics Journal*. 1988; (31)5: pp. 351-354.
6. Vysotskii V I, Smirnov I V, Komilova A A. Introduction to the Biophysics of Activated Water. Universal Publishers. Boca Raton, FLA. 2005.
7. Berg J. M., Tymoczko J. L., Stryer L. Biochemistry 5th Edition. WH Freeman. New York, 2002.
8. Vysotskii V I, Smirnov I V, Komilova A A. Introduction to the Biophysics of Activated Water. Universal Publishers. Boca Raton, FLA. 2005; p.134.
9. Vysotskii V I, Smirnov I V, Komilova A A. Introduction to the Biophysics of Activated Water. Universal Publishers. Boca Raton, FLA. 2005.
10. Salem A. A., Mossa H A, Barsoum B N. Application of nuclear magnetic resonance spectroscopy for quantitative analysis of miconazole, metronidazole and sulfamethoxazole in pharmaceutical and urine samples. J of Pharmaceutical and Biomedical Analysis. 2006; 41(2): pp. 654-661.
11. Jhon M S. The Water Puzzle and the Hexagonal Key. Uplifting Press Inc. U.S. 2004.

12. Smirnov I V. Activated Water. *Electric Spacecraft Journal*. 2002; 33: pp.15-17.
13. Smirnov I. Activated Water. *Electronic Journal of Biotechnology*. 2003; (6)2: pp.128-142.
14. Vysotskii V I, Smirnov I V, Komilova A A. Introduction to the Biophysics of Activated Water. Universal Publishers. Boca Raton, FLA. 2005; p.138.
15. Smirnov I. Activated Water. *Electronic Journal of Biotechnology*. 2003; (6)2: pp.128-142.
16. Vysotskii V I. Investigation of Physical Properties of MRET Activated Water and its Successful Application for Prophylaxis and Treatment of Oncology. Programme and Abstract Book. International Congress on Medical Physics and Biomedical Engineering, August 27 - September 1 of 2006. Seoul, Korea. 2006.
17. Smirnov I.V. The Anomalous Low Viscosity and Polarized-Oriented Multilayer Structure of MRET Activated Water. *Explore Magazine*. 2007; (16)4: pp. 37-39.
18. Smirnov I V, Peerayot T. The Physiological Effect of MRET Activated Water. *Explore Magazine*. 2006; (15)1: pp. 38-44.
19. Smirnov I V. The Effect of MRET Activated Water on Microbiological Culture Escherichia coli K-12 and on Complex Microbiological Associations. *Explore Magazine*. 2008; 17(1): pp.1-6.
20. Vysotskii V I, Smirnov I V, Kornilova A A. Introduction to the Biophysics of Activated Water. Universal Publishers. Boca Raton, FLA. 2005.
21. Vysotskii V I. Experimental Observation and the Biophysical Model of Strong Germicidal Activity of Water Activated with the help of MRET Process and Investigation of Physical Properties of MRET Activated Water and its Successful Application for Prophylaxis and Treatment of Oncology. Programme and Abstract Book, International Congress on Medical Physics and Biomedical Engineering, August 27-September 1, 2006. Seoul, Korea.

Index

GYAN.